Android Application Development

A Beginner's Tutorial

Budi Kurniawan

About the Author

Known for his clear writing style, Budi Kurniawan is a senior developer at Brainy Software and the author of *How Tomcat Works*, *Servlet and JSP: A Tutorial*, *Struts 2 Design and Programming*, and others. He has written software that is licensed by major corporations worldwide.

Table of Contents

Introduction

This book is for you if you want to learn Android application development for smart phones and tablets. Android is the most popular mobile platform today and it comes with a comprehensive set of APIs that make it easy for developers to write, test and deploy apps. With these APIs you can easily show user interface (UI) components, play and record audio and video, create games and animation, store and retrieve data, search the Internet, and so on.

The software development kit (SDK) for Android application development is free and includes an emulator, a computer program that can be configured to mimic a hardware device. This means, you can develop, debug and test your applications without physical devices.

This introduction provides an overview of the Android platform and the contents of the book.

Overview

The Android operating system is a multi-user Linux system. Each application runs as a different user in a separate Linux process. As such, an application runs in isolation from other apps.

One of the reasons for Android's rapid ascent to the top is the fact that it uses Java as its programming language. But, is Android really Java? The answer is yes and no. Yes, Java is the default programming language for Android application development. No, Android applications do not run on a Java Virtual Machine as all Java applications do. Instead, up to Android version 4.4 all Android applications run on a virtual machine called Dalvik. In version 5.0 and later, Android sources are ultimately compiled to machine code and applications run with a new runtime called ART (Android Runtime). Android 4.4 was the turning point and shipped with both Dalvik and ART.

As for the development process, initially code written in Java is compiled to Java bytecode. The bytecode is then cross-compiled to a dex (Dalvik executable) file that contains one or multiple Java classes. The dex file, resource files and other files are then packaged using the apkbuilder tool into an apk file, which is basically a zip file that can be extracted using unzip or Winzip. APK, by the way, stands for application package.

The apk file is how you deploy your app. Anyone who gets a copy of it can install and run it on his or her Android device.

In pre-5.0 versions of Android, the apk file run on Dalvik. In version 5.0 and later, the dex file in the apk is converted into machine code when the application is installed. The machine code is executed when the user runs the application. All of this is transparent to the developer and you do not have to understand intimately the dex format or the internal working of the runtime.

An apk file can run on a physical device or the emulator. Deploying an Android application is easy. You can make the apk file available for download and download it with an Android device to install it. You can also email the apk file to yourself and open the email on an Android device and install it. To publish your application on Google Play, however, you need to sign the apk file using the jarsigner tool. Fortunately, signing an apk is easy with an integrated development environment (IDE), either it is Android Studio or ADT Eclipse.

If you're interested in learning more about the Android build process, this web page explains the Android build process in detail.

```
https://developer.android.com/tools/building/index.html
```

Application Development in Brief

Before you embark on a long journey to becoming a professional Android application developer, you should know what lies ahead.

Before starting a project, you should already have an idea what Android devices will be your target. Most applications will target smart phones and tablets. However, the current Android release also allows you to develop apps for smart TVs and wearables. This book, however, is focused on application development for smart phones and tablets.

Then, you need to decide what versions of Android you want to support. Android was released in 2008, but at the time of writing this book there are already 21 API levels available, level 1 to level 21. Of course, the higher the level, the more features are available. However, many older phones and tablets do not run the latest Android and cannot run applications that target higher API levels than what are installed. For example, if you're using features in API level 21, your application will not run in Android devices that support API level 20, let alone API level 2. Fortunately, Android is backward-compatible. Applications written for an earlier version will always run on newer versions. In other words, if you write applications using API level 10, your applications will work in devices that support API level 10 and later. Therefore, you would want to aim the lowest API level possible. This topic will be discussed further in the section to come.

Once you decide what Android devices to target and the API level you should write your program in, you can start looking at the API. There are four types of Android application components:

- Activity: A window that contains user interface components.
- Service: A long running operation that runs in the background.
- Broadcast receiver: A listener that responds to a system or application announcement.
- Content provider: A component that manages a set of data to be shared with other applications.

An application can contain multiple component types, even though a beginner would normally start with an application that has one or two activities. You can think of an activity as a window. You can use Android user interface components or controls to decorate an activity and as a way to interact with the user. If you are using an IDE, you can design an activity by simply dragging and dropping controls around your computer screen.

To encourage code reuse, an application component can be offered to other applications. In fact, you should take advantage of this sharing mechanism to speed up development. For instance, instead of writing your own photo capture component, you can utilize the component of the default Camera application. Instead of writing an email sending component and reinventing the wheel, you can use the system's email application to send emails from your app.

Another important concept in Android programming is the intent. An intent is a message sent to the system or another application to request that an action be performed. You can do a lot of different things with intents, but generally you use an intent to start an activity, start a service or send a broadcast.

Every application must have a manifest, which describes the application. The manifest takes the form of an XML file and contains one or several of the following:

- The minimum API level required to run the application.
- The name of the application. This name will be displayed on the device.
- The first activity (window) that will be opened when the user touches the application icon on the Home screen of his or her phone or tablet.
- Whether or not you allow your application components be invoked from other applications. To promote code reuse, functionality in an application can be invoked from other applications as long as the author of the application agree to share it. For instance, the default Camera application can be invoked from other applications that need photo or video capture functionality.
- What set of permissions the user must grant for the application to be installed on the target device. If the user does not grant all the required permissions, the application will not install.

Yes, many things require user permissions. For example, if your application needs to store data in external storage or access the Internet, the application must request the user's permission before it can be installed. If the application needs to be automatically started when the device boots up, there is a permission for that too. In fact, there are more than 150 permissions that an application may require before it can be installed on an Android device.

Most applications are probably simple enough to only need activities and not other types of application components. Even with only activities, there is a lot to learn: UI controls, events and listeners, fragments, animation, multi-threading, graphic and bitmap processing and so on. Once you master these, you may want to look at services, broadcast receivers and content providers. All are explained in this book.

Android Versions

First released in September 2008, Android is now a stable and mature platform. The current version, version 5.0, is the 21st Android API level ever released. Table I.1 shows the code name, API level and release date of all Android *major* releases.

Version	Code Name	API Level	Release Date
1.0		1	September 23, 2008
1.1		2	February 9, 2009
1.5	Cupcake	3	April 30, 2009
1.6	Donut	4	September 15, 2009
2.0	Eclair	5	October 26, 2009
2.0.1	Eclair	6	December 3, 2009
2.1	Eclair	7	January 12, 2010
2.2	Froyo	8	May 20, 2010
2.3	Gingerbread	9	December 6, 2010
2.3.3	Gingerbread	10	February 9, 2011
3.0	Honeycomb	11	February 22, 2011
3.1	Honeycomb	12	May 10, 2011
3.2	Honeycomb	13	July 15, 2011
4.0	Ice Cream Sandwich	14	October 19, 2011
4.0.3	Ice Cream Sandwich	15	December 16, 2011
4.1	Jelly Bean	16	July 9, 2012
4.2	Jelly Bean	17	November 13, 2012
4.3	Jelly Bean	18	July 24, 2013
4.4	Kitkat	19	October 31, 2013
4.4w	Kitkat	20	July 22, 2014
5.0	Lollipop	21	November 3, 2014

Table I.1: Android versions

Note

Version 4.4w is the same as 4.4 but with wearable extensions.

With each new version, new features are added. As such, you can use the most features if you target the latest Android release. However, not every Android phone and tablet is running the latest release because Android devices made for older APIs may not support later releases and software upgrade is not always automatic. Table I.2 shows Android versions still in use today.

Version	Codename	API	Distribution
2.2	Froyo	8	0.5%
2.3.3-2.3.7	Gingerbread	10	9.1%
4.0.3-4.0.4	Ice Cream Sandwich	15	7.8%
4.1.x	Jelly Bean	16	21.3%
4.2.x		17	20.4%
4.3		18	7.0%
4.4	KitKat	19	33.9%

Table I.2: Android versions still in use (December 2014)

The data in Table I.2 was taken from this web page:

```
https://developer.android.com/about/dashboards/index.html
```

If you distribute your application through Google Play, the most popular marketplace for Android applications, the lowest version of Android that can download your application is 2.2, because versions older than 2.2 cannot access Google Play. In general you would want to reach as wide customer base as possible, which means supporting version 2.2 and up. If you only support version 4.0 and up, for example, you leave out 9.6% of Android devices, which may or may not be okay.

However, the lower the version, the fewer features are supported. Some people risk alienating some customers in order to be able to use the more recent features. To alleviate this problem, Google provides a support library that allows you to use newer features in old devices which otherwise would not be able to enjoy those features. You will learn how to use this support library in this book.

Online Reference

The first challenge facing new Android programmers is understanding the components available in Android. Luckily, documentation is in abundance and it is easy to find help over the Internet. The documentation of all Android classes and interfaces can be found on Android's official website:

```
http://developer.android.com/reference/packages.html
```

Undoubtedly, you will frequent this website as long as you work with Android. If you had a chance to browse the website, you'd have learned that the first batch of types belong to the **android** package and its subpackages. After them come the **java** and **javax** packages that you can use in Android applications. Java packages that cannot be used in Android, such as **javax.swing** and **java.nio.file**, are not listed there.

Which Java Versions Can I Use?

To develop Android applications using tools such as Android Studio or ADT Eclipse, you need JDK 6 or later. Support for Java 7 language features was added to Android 4.4 (Kitkat).

At the time of writing there is no official support for Java 8 yet. However, if you are using Android Studio and have JDK 8 installed, you can already use lambda expressions, a new major feature in Java 8.

About This Book

This section outlines the contents of this book.

Chapter 1, "Getting Started" shows how to install the Android SDK and Android Studio and create a simple Android application.

Chapter 2, "Activities" explains the activity and its lifecycle. The activity is one of the most important concepts in Android programming.

Chapter 3, "UI Components" covers the more important UI components, including widgets, Toast, AlertDialog and notifications.

Chapter 4, "Layouts" shows how to lay out UI components in Android applications and use the built-in layouts available in Android.

Chapter 5, "Listeners" talks about creating a listener to handle events.

Chapter 6, "The Action Bar" shows how you can add items to the action bar and use it to drive application navigation.

Menus are a common feature in many graphical user interface (GUI) systems whose primary role is to provide shortcuts to certain actions. Chapter 7, "Menus" looks at Android menus closely.

Chapter 8, "ListView" explains about **ListView**, a view that shows a scrollable list of items and gets its data source from a list adapter.

Chapter 9, "GridView" covers the **GridView** widget, a view similar to the **ListView**. Unlike the **ListView**, however, the **GridView** displays its items in a grid.

Chapter 10, "Styles and Themes" discusses the two important topics directly responsible for the look and feel of your apps.

Chapter 11, "Bitmap Processing" teaches you how to manipulate bitmap images. The techniques discussed in this chapter are useful even if you are not writing an image editor application.

The Android SDK comes with a wide range of views that you can use in your applications. If none of these suits your need, you can create a custom view and draw on it. Chapter 12, "Graphics and Custom Views" shows how to create a custom view and draw shapes on a canvas.

Chapter 13, "Fragments" discusses fragments, which are components that can be added to an activity. A fragment has its own lifecycle and has methods that get called when certain phases of its life occur.

Chapter 14, "Multi-Pane Layouts" shows how you can use different layouts for different screen sizes, like that of handsets and that of tablets.

Chapter 15, "Animation" discusses the latest Animation API in Android called property animation. It also provides an example.

Chapter 16, "Preferences" teaches you how to use the Preference API to store application settings and read them back.

Chapter 17, "Working with Files" show how to use the Java File API in an Android application.

Chapter 18, "Working with the Database" discusses the Android Database API, which you can use to connect to an SQLite database. SQLite is the default relational database that comes with every Android device.

Chapter 19, "Taking Pictures" teaches you how to take still images using the built-in Camera application and the Camera API.

Chapter 20, "Making Videos" shows the two methods for providing video-making capability in your application, by using a built-in intent or by using the **MediaRecorder** class.

Chapter 21, "The Sound Recorder" shows how you can record audio.

Chapter 22, "Handling the Handler" talks about the **Handler** class, which can be used, among others, to schedule a **Runnable** at a future time.

Chapter 23, "Asynchronous Tasks" explains how to handle asynchronous tasks in Android.

Chapter 24, "Services" explains how to create background services that will run even after the application that started them was terminated.

Chapter 25, "Broadcast Receivers" discusses another kind of android component for receiving intent broadcasts.

Chapter 26, "The Alarm Manager" shows how you can use the **AlarmManager** to schedule jobs.

Chapter 27, "Content Providers" explains yet another application component type for encapsulating data that can be shared across applications.

Appendix A, "Installing the JDK" provides instructions on how to install the JDK.

Appendix B, "Using the ADT Bundle" explains how to use the ADT Bundle as an alternative IDE to develop Android apps.

Code Download

The examples accompanying this book can be downloaded from the publisher's website:

```
http://books.brainysoftware.com
```

Chapter 1
Getting Started

You need the Android Software Development Kit (SDK) to develop, debug and test your applications. The SDK contains various tools including an emulator to help you test your applications without a physical device. Currently the SDK is available for Windows, Mac OS X and Linux operating systems.

You also need an integrated development environment (IDE) to speed up development. You could build applications without an IDE, but that would be more difficult and unwise. There are two IDEs currently available, both free:

- Android Studio, which is based on IntelliJ IDEA, a popular Java IDE. This software suite includes the Android SDK.
- The Android Developer Tools (ADT) Bundle, a bundle that includes the Android SDK and Eclipse. Eclipse is another popular Java IDE.

Released in December 2014, Android Studio is the preferred IDE and the ADT bundle will not be supported in the future. Therefore, you should start using Android Studio unless you have very good reasons to choose the ADT Bundle. This book assumes you are using Android Studio.

In this chapter you will learn how to download and install Android Studio. After you have successfully installed the IDE, you will write and build your first Android application and run it on the emulator.

Android application development requires a Java Development Kit (JDK). For Android 5 or later, or if you are developing using Android Studio, you need JDK 7 or later. For pre-5 Android, you need JDK 6 or later. If you do not have a JDK installed, make sure you download and install one by following the instructions in Appendix A, "Installing the JDK."

Downloading and Installing Android Studio

You can download Android Studio from this web page:

```
http://developer.android.com/sdk/index.html
```

Android Studio is available for Windows, Mac OS X and Linux. Installing Android Studio also downloads and installs the Android SDK.

Installing on Windows

Follow these steps to install Android Studio on Windows.

1. Double-click the exe file you downloaded to launch the Setup wizard. The welcome page of the wizard is shown in Figure 1.1.

Figure 1.1: The Android Studio Setup program

2. Click **Next** to proceed.

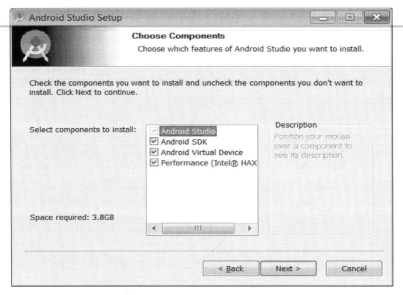

Figure 1.2: Choosing components

3. You will see the next dialog of the Setup wizard as shown in Figure 1.2. Here you can choose the components do install. Leave all components selected and click **Next** again.

Figure 1.3: The license agreement

4. The next dialog, shown in Figure 1.3, shows the license agreement. You really have no choice but to agree on the license agreement if you wish to use Android Studio, in which case you have to click **I Agree**.

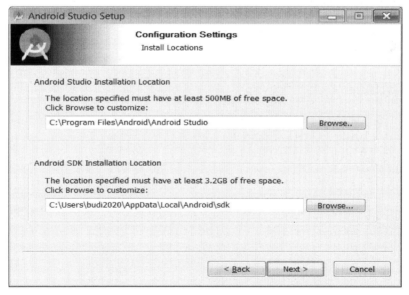

Figure 1.4: Choosing the install locations

5. In the next dialog that appears, which is shown in Figure 1.4, browse to the install locations for both Android Studio and the Android SDK. Android Studio should come with suggestions. It's not a bad idea to accept the locations suggested. Once you find locations for the software, click **Next**.

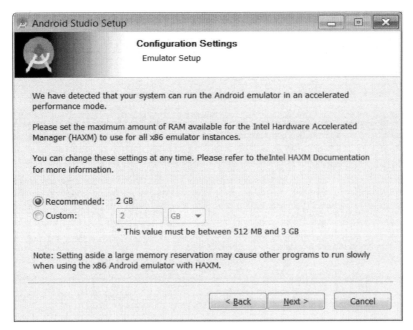

Figure 1.5: Emulator setup

6. The next dialog, presented in Figure 1.5, shows the configuration page for the emulator. Click **Next**.

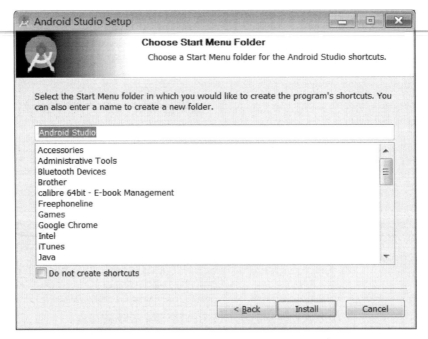

Figure 1.6: Choosing the Start menu folder

7. The next dialog, shown in Figure 1.6, is the last dialog before installation. Here you have to select a Start menu folder. Simply accept the default value and click

Install. Android Studio will start to install.

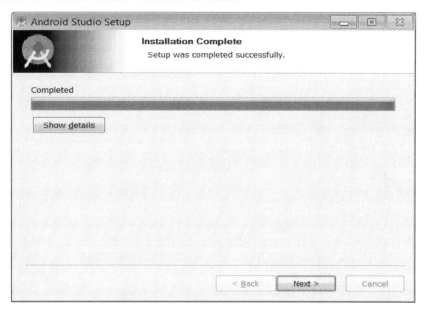

Figure 1.7: Installation complete

8. Once installation is complete, you will see another dialog similar to that in Figure 1.7. Click **Next**.

Figure 1.8: Setup is finished

9. On the next dialog, shown in Figure 1.8, click **Finish**, leaving the "Start Android Studio" checkbox checked. If you have installed a previous version of Android Studio, the Setup wizard will ask you if you want to import settings from the

previous version of Android Studio. This is shown in Figure 1.9.

Figure 1.9: Deciding whether to import settings from another version of Android Studio

10. Leave the second radio button checked and click **OK**. The Setup wizard will quietly create an Android virtual device and report it to you once it's finished. (See Figure 1.10).

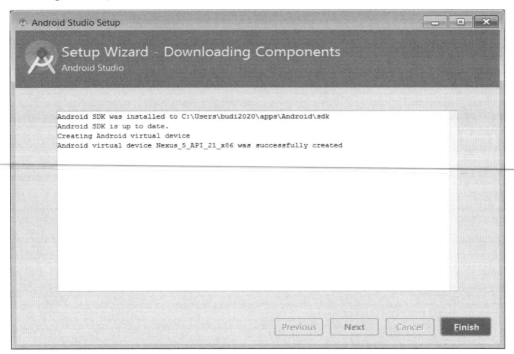

Figure 1.10: The Setup wizard has just created an AVD

11. Click **Finish**. Finally, Android Studio is ready to use. The welcome dialog is shown in Figure 1.11.

Figure 1.11: Android Studio's welcome dialog

Installing on Mac OS X

To install Android Studio on a Mac OS X machine, follow these steps:

1. Launch the dmg file you downloaded.
2. Drag and drop Android Studio to the Applications folder.
3. Open Android Studio and follow the setup wizard to install the SDK.

Installing on Linux

On Linux, extract the downloaded zip file, open a terminal and change directory to the **bin** directory of the installation directory and type:

```
./studio.sh
```

Then, follow the setup wizard to install the SDK.

Creating An Application

Creating an Android application with Android Studio is as easy as a few mouse clicks. This section shows how to create a Hello World application, package it, and run it on the emulator. Make sure you have installed the Android SDK and Android Studio by following the instructions in the previous section.

Next, follow these steps.

1. Click the **File** menu in Android Studio and select **New Project**. The first dialog of

the **Create New Project** wizard, shown in Figure 1.12, appears.

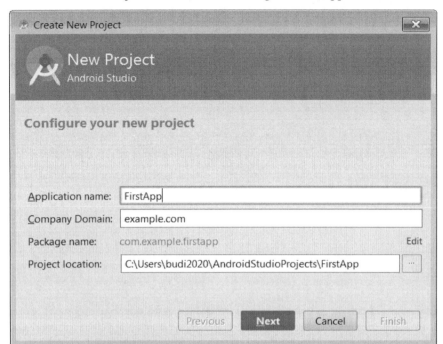

Figure 1.12: Entering application details

2. Enter the details of the new application. In the **Application name** field, type the name to appear on the Android device. In the **Company Domain** field, type your company's domain. If you do not have one, just use **example.com**. The company domain in reverse order will be used as the base package name for the application. The package name uniquely identifies your application. You can change the package name by clicking the **Edit** button to the right of the field. By default, the project will be created under the **AndroidStudioProjects** directory created when you installed Android Studio. You can change the location too if you wish.

3. Click **Next**. The second dialog opens as shown in Figure 1.13. Here you need to select a target (phone and Tablet, TV, etc) and the minimum API level. This book only discusses Android application development for phones and tablets, so leave the selected option checked. As for the minimum API level, the lower the level, the more devices your application can run on, but the fewer features are available to you. For now, keep the API level Android Studio has selected for you.

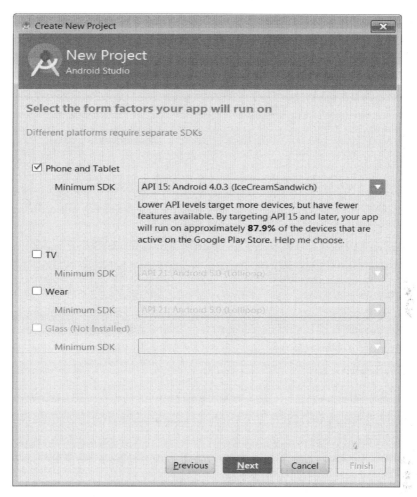

Figure 1.13: Selecting a target

4. Click **Next** again. A dialog similar to that in Figure 1.14 appears. Android Studio is asking you if you want to add an activity to your project and, if so, what kind of activity. At this stage, you probably do not know what an activity is. For now, think of it as a window, and add a blank activity to your project. So, accept the selected activity type.

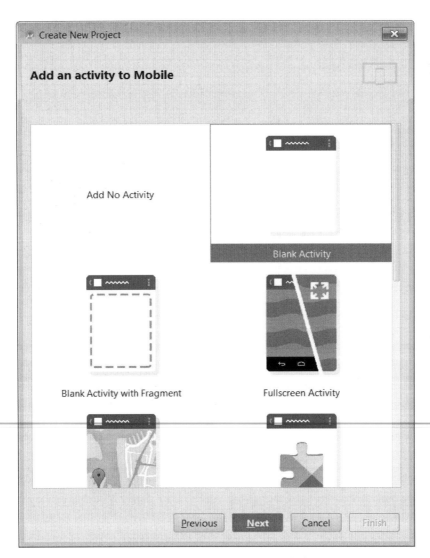

Figure 1.14: Adding an activity

5. Click **Next** again. The next dialog that appears looks like the dialog in Figure 1.15. In this dialog you can enter a Java class name for your activity class as well as a title for your activity window and a layout name. For now just accept the default.

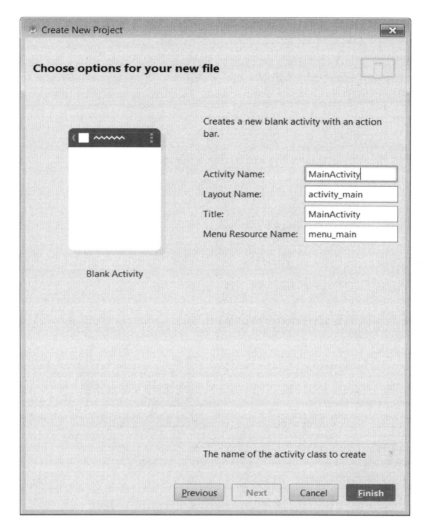

Figure 1.15: Entering the activity class name and other details

5. Click **Finish**. Android Studio will prepare your project and it may take a while. Finally, when it's finished, you will see your project in Android Studio, like the one shown in Figure 1.16.

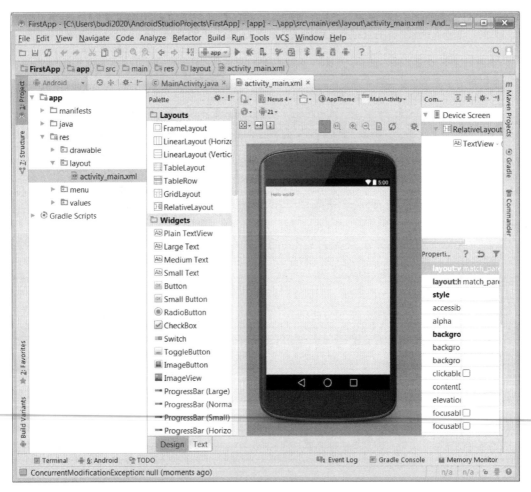

Figure 1.16: The new Android project

The next section shows you how you can run your application on the emulator.

Running the Application on the Emulator

Now that you have an application ready, you can run it by clicking the Run button. You will be asked to choose a device.

Figure 1.17: Selecting a device to run the application

If you have not created an emulator, do so now. If you have, you will see all running emulators. Or, you can launch one. Click "Use same device for future launches" to use the same emulator in the future.

Next, click **OK**.

It will take seconds to launch the AVD. As you know, the emulator emulates an Android device. Just like a physical device, you need to unlock the emulator's screen when running your app for the first time.

If your application does not open automatically, locate the application icon and double-click on it. Figure 1.18 shows the application you just created.

Figure 1.18: Your application running on the emulator

During development, leave the emulator running while you edit your code. This way, the emulator does not need to be loaded again every time you test your application.

The Application Structure

Now, after the little excitement of having just run your first Android application, let's go back to Android Studio and take a look at the structure of an Android application. Figure 1.19 shows the left treeview that contains the project components.

Figure 1.19: The application structure

There are two main nodes in the Project window in Android Studio, **app** and **Gradle Scripts**. The **app** node contains all the components in the application. The **Gradle Scripts** node contains the Gradle build scripts used by Android Studio to build your project. I will not discuss these scripts, but it would be a good idea for you to get familiar with Gradle.

There are three nodes under the **app** node:

- **manifests**. Contains an **AndroidManifest.xml** file that describes your application. It will be explained in more detail in the next section "The Android Manifest."
- **java**. Contains all Java application and test classes.
- **res**. Contains resource files. Underneath this directory are these directories: **drawable** (containing images for various screen resolutions), **layout** (containing layout files), **menu** (containing menu files) and **values** (containing string and other values).

The R Class

Not visible from inside Android Studio is a generated Java class named **R**, which can be found in the **app/build/generated/source** directory of the project. **R** contains nested classes that in turn contain all the resource IDs for all your resources. Every time you add, change or delete a resource, **R** is re-generated. For instance, if you add an image file named **logo.png** to the **res/drawable** directory, Android Studio will generate a field named **logo** under the **drawable** class, a nested class in **R**.

The purpose of having **R** is so that you can refer to a resource in your code. For instance, you can refer to the **logo.png** image file with **R.drawable.logo**.

The Android Manifest

Every Android application must have a manifest file called **AndroidManifest.xml** file that describes the application. Listing 1.1 shows a sample manifest file.

Listing 1.1: A sample manifest

```xml
<?xml version="1.0" encoding="utf-8"?>
<manifest xmlns:android="http://schemas.android.com/apk/res/android"
    package="com.example.firstapp" >

    <application
        android:allowBackup="true"
        android:icon="@drawable/ic_launcher"
        android:label="@string/app_name"
        android:theme="@style/AppTheme" >
        <activity
            android:name="com.example.firstapp.MainActivity"
            android:label="@string/app_name" >
            <intent-filter>
                <action android:name="android.intent.action.MAIN" />
                <category
android:name="android.intent.category.LAUNCHER" />
            </intent-filter>
        </activity>
    </application>
</manifest>
```

A manifest file is an XML document with **manifest** as the root element. The **package** attribute of the **manifest** element specifies a unique identifier for the application. Android tools will also use this information to generate appropriate Java classes that are used from the Java source you write.

Under **<manifest>** is an **application** element that describes the application. Among others, it contains one or more **activity** elements that describe activities in your app. An application typically has a main activity that serves as the entry point to the application. The **name** attribute of an **activity** element specifies an activity class. It can be a fully qualified name or just the class name. If it is the latter, the class is assumed to be in the package specified by the **package** attribute of the **manifest** element. In other words, the **name** attribute of the above **activity** element can be written as one of the following:

```
android:name="MainActivity"
```

```
android:name=".MainActivity"
```

You can reference a resource from your manifest file (and other XML files in the project) using this format:

```
@resourceType/name
```

For example, these are some of the attributes of the application element in Listing 1.1:

```
android:icon="@drawable/ic_launcher"
android:label="@string/app_name"
android:theme="@style/AppTheme"
```

The first attribute, **android:icon**, refers to a drawable named **ic_launcher**. If you browse the project in Android Studio, you can find an **ic_launcher.png** file under **res/drawable**.

The second attribute, **android:label**, refers to a string resource called **app_name**. All string resources are located in the **strings.xml** file under **res/values**.

Finally, the third attribute, **android:theme**, references a style named **AppTheme**. All styles are defined in the **styles.xml** file under **res/values**. Styles and themes are discussed in Chapter 10, "Styles and Themes."

There are other elements that may appear in the Android manifest and you will learn to use many of them in this book. You can find the complete list of elements here:

```
http://developer.android.com/guide/topics/manifest/manifest-
element.html
```

The APK File

An Android application is packaged into an apk file, which is basically a zip file and can be opened using WinZip or a similar program. All applications are signed with a private key. This process sounds hard enough, but thankfully Android Studio takes care of everything. When you run an Android application from inside Android Studio, an apk file will be built and signed automatically. The file will be named **app-debug.apk** and stored in the **app/build/outputs/apk** directory under the project directory. Android Studio also notifies the emulator or the target device of the location so that the apk file can be found and executed.

The automatically generated apk file also contains debug information to enable running it in debug mode.

Figure 1.20 shows the structure of the apk file that is created when you run your application.

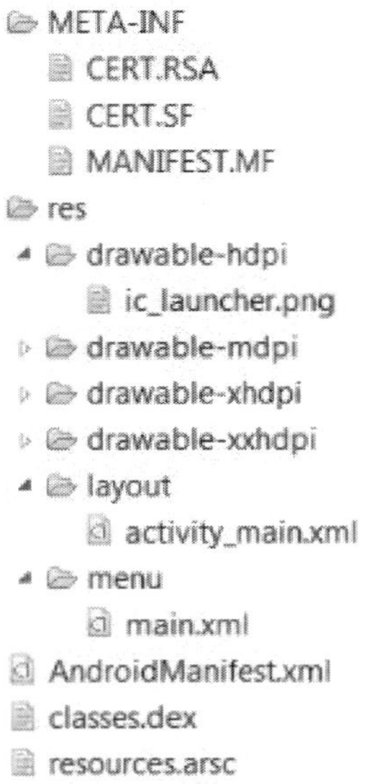

```
META-INF
    CERT.RSA
    CERT.SF
    MANIFEST.MF
res
    drawable-hdpi
        ic_launcher.png
    drawable-mdpi
    drawable-xhdpi
    drawable-xxhdpi
    layout
        activity_main.xml
    menu
        main.xml
AndroidManifest.xml
classes.dex
resources.arsc
```

Figure 1.20: Android application structure

The manifest file is there and so are the resource files. The **AndroidManifest.xml** file is compiled so you cannot use a text editor to read it. There is also a **classes.dex** file that contains the binary translation of your Java classes into Dalvik executable. Note that even if you have multiple java files in your application, there is only one **classes.dex** file.

Debuging Your Application

Android Studio is full of useful features for rapidly developing and testing your application. One of the features is support for debugging.

The following are some of the ways you can debug your application.

Logging

The easiest way to debug an application is by logging messages. Java programmers like to use logging utilities, such as Commons Logging and Log4J, to log messages. The Android framework provides the **android.util.Log** class for the same purpose. The **Log** class comes with methods to log messages at different log levels. The method names are short: **d** (debug), **i** (info), **v** (verbose), **w** (warning), **e** (error), and **wtf** (what a terrible failure).

This methods allow you to write a tag and the text. For example,

```
Log.e("activity", "Something went wrong");
```

During development, you can watch the Android DDMS view at the bottom of Android Studio's main screen.

The good thing about LogCat is that messages at different log levels are displayed in different colors. In addition, each message has a tag and this makes it easy to find a message. In addition, LogCat allows you to save messages to a file and filter the messages so only messages of interest to you are visible.

The LogCat view is shown in Figure 1.21.

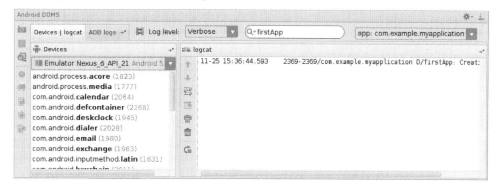

Figure 1.21: LogCat in Android DDMS

Any runtime exception thrown, including the stack trace, will also appear in LogCat, so you can easily identify which line of code is causing the problem.

Setting Breakpoints

The easiest way to debug an application is by logging messages. However, if this does not help and you need to trace your application, you can use other debugging tools in Android Studio.

Try adding a line breakpoint in your code by clicking on a line and selecting **Run > Toggle Line Breakpoint**. Figure 1.22 shows a line breakpoint in the code editor.

```java
    };
    IntentFilter intentFilter = new IntentFilter(Intent.ACTION_TIME_TICK);

    @Override
    protected void onCreate(Bundle savedInstanceState) {
        super.onCreate(savedInstanceState);
        setContentView(R.layout.activity_main);
        Log.d("firstApp", "Creating activity");
    }

    @Override
    protected void onResume() {
        this.registerReceiver(timeTickReceiver, intentFilter);
        super.onResume();
    }

    @Override
    protected void onPause() {
        this.unregisterReceiver(timeTickReceiver);
        super.onPause();
    }
    @Override
    public boolean onCreateOptionsMenu(Menu menu) {
        getMenuInflater().inflate(R.menu.menu_main, menu);
        return true;
    }
}
```

Figure 1.22: Setting a line breakpoint

Now, debug your application by selecting **Run** > **Debug app**.

The Debug view is shown in Figure 1.23.

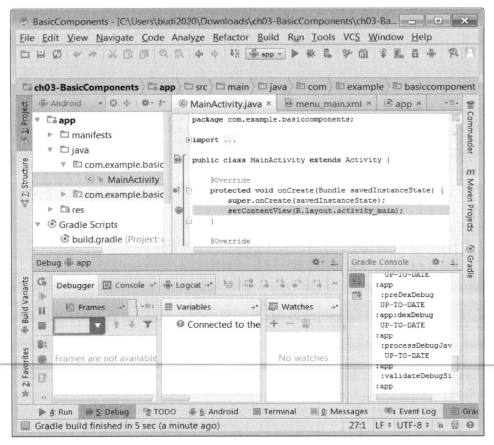

Figure 1.23: The Debug view

Here, you can step into your code, view variables, and so on.

The Android SDK Manager

When you install Android Studio, the setup program also downloads the latest version of the Android SDK. You can manage the packages in the Android SDK using the Android SDK Manager.

To launch the Android SDK Manager, in Android Studio click **Tools** > **Android** > **SDK Manager**. Alternatively, click the SDK Manager button on the toolbar.

The SDK Manager button is shown in Figure 1.24.

Figure 1.24: The SDK Manager button on the toolbar

The SDK Manager window is shown in Figure 1.25.

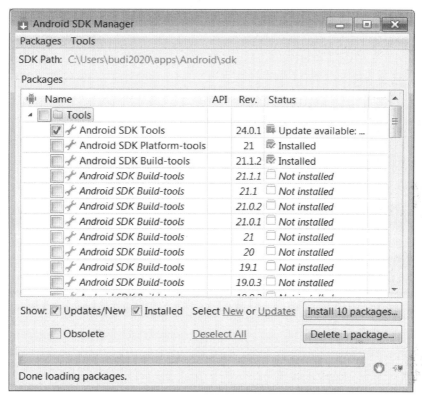

Figure 1.25: The Android SDK Manager window

In the Android SDK Manager, you can download other versions of the SDK or delete components you do not need.

Creating An Android Virtual Device

The SDK ships with an emulator so that you can test your applications without a physical device. The emulator can be configured to mimic various Android phones and tablets, from Nexus S to Nexus 9. Each instance of the configured emulator is called an Android virtual device (AVD). You can create multiple virtual devices and run them simultaneously to test your application in multiple devices.

When you install Android Studio, it also creates an Android virtual device. You can create more virtual devices using the Android Virtual Device (AVD) Manager.

To create an AVD, open the Android Virtual Device (AVD) Manager. You can open it by clicking **Tools** > **Android** > **AVD Manager**. Alternatively, simply click the AVD Manager button on the toolbar. Figure 1.26 shows the AVD Manager button

Figure 1.26: The AVD Manager button on the toolbar

If you have not created a single AVD in your machine, the first window of the AVD Manager will look like that in Figure 1.27. If you have created virtual devices before, the first window will list all the devices.

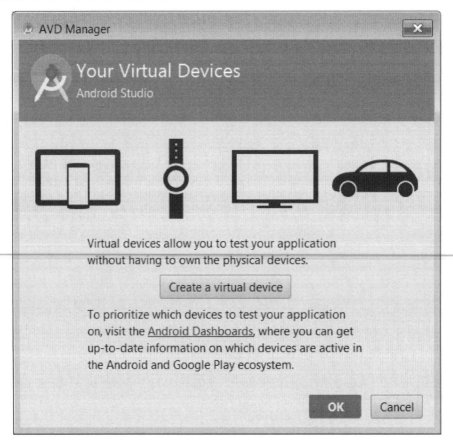

Figure 1.27: The AVD Manager's welcome screen

To create an AVD, follow these steps.

1. Click the **Create a virtual device** button. You will see a window similar to that in Figure 1.28.

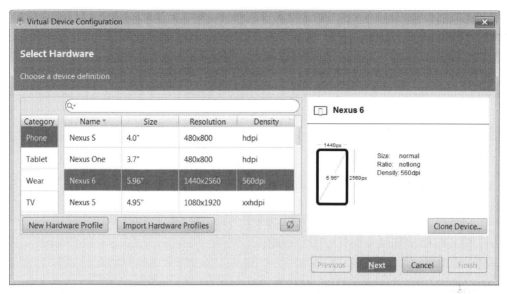

Figure 1.28: Selecting a phone profile

2. Select **Phone** from Category and then select a device from the center window. Next, click the **Next** button. The next window will show. See Figure 1.29.

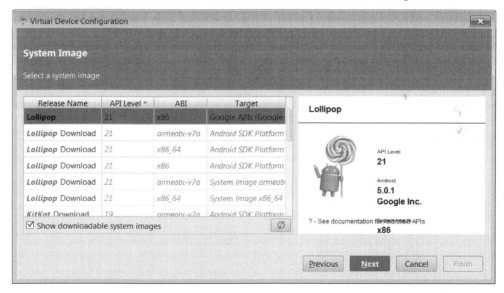

Figure 1.29: Selecting the API level and ABI

3. Select an API level and application binary interface (ABI). If you are using a 32-bit Intel CPU, then it must be x86. If it is a 64-bit Intel CPU, chances are you need the x86_64.
4. Click the **Next** button. In the next step you will be asked to verify the configuration details of the AVD you are creating. (See Figure 1.30.)

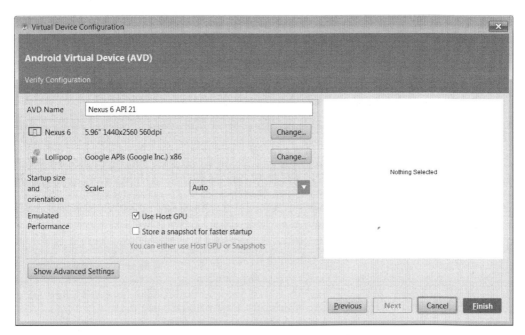

Figure 1.30: Verifying the details of the AVD

5. Click the **Finish** button. It will take more than a few seconds for the AVD Manager to create a new emulator. Once it's finished, you will see a list like the one in Figure 1.31.

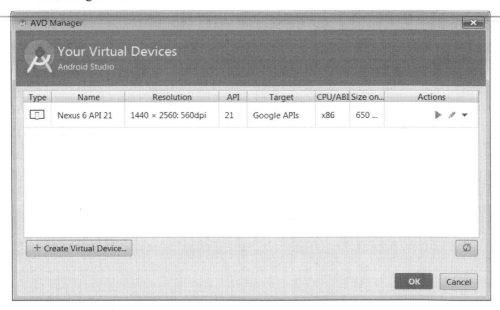

Figure 1.31: A list of available AVDs

For each AVD, there are three action buttons in the rightmost column. The first icon, a green arrow, is for launching the emulator. The second, a pencil, is for editing the emulator details. The last one, a down arrow, shows more actions such as Delete and View Details.

Running An Application on A Physical Device

There are a couple of reasons for wanting to test your application on a real device. The most compelling one is that you should test your applications on real devices before publishing them. Other reasons include speed. An emulator may not be as fast as a new Android device. Also, it is not always easy to simulate certain user inputs in an emulator. For example, you can change the screen orientation easily with a real device. On the emulator, you have to press **Ctrl+F12**.

To run your application on a real device, follow these steps.

1. Declare your application as debuggable by adding **android:debuggable="true"** in the **application** element in the manifest file.
2. Enable USB debugging on the device. On Android 3.2 or older, the option is under **Settings > Applications > Development**. On Android 4.0 and later, the option is under **Settings > Developer Options**. On Android 4.2 and later, Developer options is hidden by default. To make it visible, go to **Settings** > **About phone** and tap **Build number** seven times.

Next, set up your system to detect the device. The step depends on what operating system you're using. For Mac users, you can skip this step. It will just work.

For Windows users, you need to install the USB driver for Android Debug Bridge (adb), a tool that lets you communicate with an emulator or connected Android device. You can find the location of the driver from this site.

```
http://developer.android.com/tools/extras/oem-usb.html
```

For Linux users, please see the instructions here.

```
http://developer.android.com/tools/device.html
```

Opening A Project in Android Studio

You can download the Android Studio projects accompanying this book from the publisher's website. To open a project, select **File > Open** and browse to the application directory. Figure 1.32 shows how the **Open File or Project** window looks like.

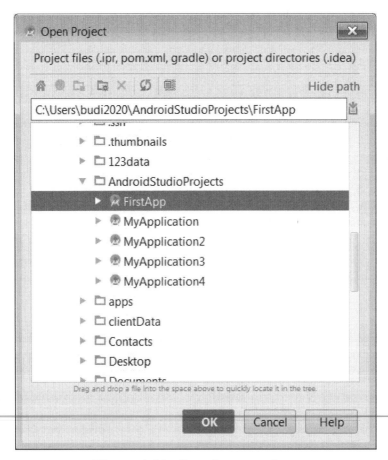

Figure 1.32: Opening a project

Using Java 8

By default, Android Studio can compile sources with Java 7 language syntax. However, you can use Java 8 language features, even though Java 8 is not yet officially supported. It goes without saying that you need JDK 8 or later to use the higher language level. Also, even though you can use the Java 8 language features, you still cannot use the libraries that come with Java 8, such as the new Date Time API or the Stream API.

If you really want to use Java 8 to write Android applications, here is how you can change the Java language level from 7 to 8 in Android Studio.

1. Expand the Gradle Scripts node on the Project view. You will see two **build.gradle** nodes on the list. Double-click the second build file to open it. You will see something like this:

```
android {
    compileSdkVersion 21
    buildToolsVersion "19.1.0"
```

```
        defaultConfig {
            ...
        }
        buildTypes {
            ...
        }
    }
```

2. Add the line in bold to the build file to change the language level to 7.

```
android {
    compileSdkVersion 21
    buildToolsVersion "19.1.0"

    defaultConfig {
        ...
    }
    buildTypes {
        ...
    }
    compileOptions {
        sourceCompatibility JavaVersion.VERSION_1_8
        targetCompatibility JavaVersion.VERSION_1_8
    }
}
```

As changing the language level adds complexity to a project, this book will stick with Java 7.

Getting Rid of the Support Library

When you create a new project with Android Studio, it structures the application to use the Android support library, so that your application can be run with a lower API level. While this might help, in many practical circumstances, you may not want the support library. Fortunately, you can remove the support library quite easily by following these steps.

1. In the app's **build.gradle** file, remove the dependency on appcompat-v7 by removing or commenting out the corresponding line:

```
dependencies {
    compile fileTree(dir: 'libs', include: ['*.jar'])
    // compile 'com.android.support:appcompat-v7:21.0.2'
}
```

2. Save the **build.gradle** file. A message in light yellow background will appear on the top part of the editor, asking you to synchronize the project. Click **Sync now**.
3. In the **res/values/styles.xml** file, assign **android:Theme.Holo** or **android:Theme.Holo.Light** to the **parent** attribute, like so

```
<style name="AppTheme" parent="android:Theme.Holo">
```

```
        <!-- Customize your theme here. -->
    </style>
```

4. Change **ActionBarActivity** in every activity class to **Activity** and remove the **import** statement that imports **ActionBarActivity**. The shortcut for this in Android Studio is **Ctrl+Alt+O**.

5. In all the **menu.xml** files, replace **app:showAsAction** with **android:showAsAction**. For example, replace

    ```
    app:showAsAction="never"
    ```
    with
    ```
    android:showAsAction="never"
    ```

6. Rebuild the project by selecting **Project > Rebuild Project**.

Summary

This chapter discusses how to install the required software and create your first application. You also learned how to create a virtual device so you can test your app in multiple devices without physical devices.

Chapter 2
Activities

In Chapter 1, "Getting Started" you learned to write a simple Android application. It is now time to delve deeper into the art and science of Android application development. This chapter discusses one of the most important component types in Android programming, the activity.

The Activity Lifecycle

The first application component that you need to get familiar with is the activity. An activity is a window containing user interface (UI) components that the user can interact with. Starting an activity often means displaying a window.

An activity is an instance of the **android.app.Activity** class. A typical Android application starts by starting an activity, which, as I said, loosely means showing a window. The first window that the application creates is called the main activity and serves as the entry point to the application. Needless to say, an Android application may contain multiple activities and you specify the main activity by declaring it in the application manifest file.

For example, the following **application** element in an Android manifest defines two activities, one of which is declared as the main activity using the **intent-filter** element. To make an activity the main activity of an application, its **intent-filter** element must contain the **MAIN** action and **LAUNCHER** category like so.

```
<application ... >
    <activity
            android:name="com.example.MainActivity"
            android:label="@string/app_name" >
        <intent-filter>
            <action android:name="android.intent.action.MAIN"/>
            <category
                android:name="android.intent.category.LAUNCHER"/>
        </intent-filter>
    </activity>
    <activity
            android:name="com.example.SecondActivity"
            android:label="@string/title_activity_second" >
    </activity>
</application>
```

In the snippet above, it is not hard to see that the first activity is the main activity.

When the user selects an application icon from the Home screen, the system will look for the main activity of the application and start it. Starting an activity entails instantiating the activity class (which is specified in the **android:name** attribute of the **activity** element in the manifest) and calling its lifecycle methods. It is important that you understand these methods so you can write code correctly.

The following are the lifecycle methods of **Activity**. Some are called once during the application lifetime, some can be called more than once.

- **onCreate**
- **onStart**
- **onResume**
- **onPause**
- **onStop**
- **onRestart**
- **onDestroy**

To truly understand how these lifecycle methods come into play, consider the diagram in Figure 2.1.

The system begins by calling the **onCreate** method to create the activity. You should place the code that constructs the UI here. Once **onCreate** is completed, your activity is said to be in the **Created** state. This method will only be called once during the activity life time.

Next, the system calls the activity's **onStart** method. When this method is called, the activity becomes visible. Once this method is completed, the activity is in the **Started** state. This method may be called more than once during the activity life time.

onStart is followed by **onResume** and once **onResume** is completed, the activity is in the **Resumed** state. How I wish they had called it **Running** instead of **Resumed**, because the fact is this is the state where your activity is fully running. **onResume** may be called multiple times during the activity life time.

Therefore, **onCreated**, **onStart**, and **onResume** will be called successively unless something goes awry during the process. Once in the **Resumed** state, the activity is basically running and will stay in this state until something occurs to change that, such as if the alarm clock sets off or the screen turns off because the device is going to sleep, or perhaps because another activity is started.

The activity that is leaving the **Resumed** state will have its **onPause** method called. Once **onPause** is completed, the activity enters the **Paused** state. **onPause** can be called multiple times during the activity life time.

What happens after **onPause** depends on whether or not your activity becomes completely invisible. If it does, the **onStop** method is called and the activity enters the **Stopped** state. On the other hand, if the activity becomes active again after **onPause**, the system calls the **onResume** method and the activity re-enters the **Resumed** state.

An activity in the **Stopped** state may be re-activated if the user chooses to go back to the activity or for some other reason it goes back to the foreground. In this case, the **onRestart** method will be called, followed by **onStart**.

Finally, when the activity is decommissioned, its **onDestroy** method is called. This method, like **onCreate**, can only be called once during the activity life time.

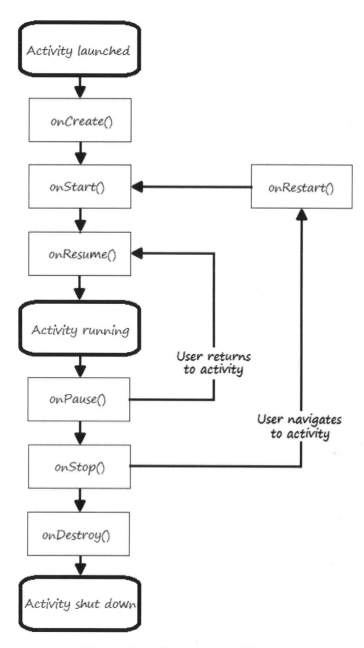

Figure 2.1: The activity lifecycle

ActivityDemo Example

The ActivityDemo application accompanying this book demonstrates when the activity lifecycle methods are called. Listing 2.1 shows the manifest for this application.

Listing 2.1: The manifest for ActivityDemo

```xml
<?xml version="1.0" encoding="utf-8"?>
<manifest xmlns:android="http://schemas.android.com/apk/res/android"
    package="com.example.activitydemo"
    android:versionCode="1"
    android:versionName="1.0" >

    <uses-sdk
        android:minSdkVersion="8"
        android:targetSdkVersion="21" />

    <application
        android:allowBackup="true"
        android:icon="@drawable/ic_launcher"
        android:label="@string/app_name"
        android:theme="@style/AppTheme" >
        <activity
            android:name="com.example.activitydemo.MainActivity"
            android:screenOrientation="landscape"
            android:label="@string/app_name" >
            <intent-filter>
                <action android:name="android.intent.action.MAIN" />
                <category
                    android:name="android.intent.category.LAUNCHER" />
            </intent-filter>
        </activity>
    </application>

</manifest>
```

This manifest is like the one in Chapter 1, "Getting Started." It has one activity, the main activity. However, notice that I specify the orientation of the activity using the **android:screenOrientation** attribute of the activity element.

The main class for this application is printed in Listing 2.2. The class overrides all the lifecycle methods of **Activity** and prints a debug message in each lifecycle method.

Listing 2.2: The MainActivity class for ActivityDemo

```java
package com.example.activitydemo;
import android.os.Bundle;
import android.app.Activity;
import android.util.Log;
import android.view.Menu;

public class MainActivity extends Activity {

    @Override
    protected void onCreate(Bundle savedInstanceState) {
        super.onCreate(savedInstanceState);
        Log.d("lifecycle", "onCreate");
        setContentView(R.layout.activity_main);
    }
```

```java
@Override
public boolean onCreateOptionsMenu(Menu menu) {
    // Inflate the menu; this adds items to the action bar
    // if it is present.
    getMenuInflater().inflate(R.menu.menu_main, menu);
    return true;
}

@Override
public void onStart() {
    super.onStart();
    Log.d("lifecycle", "onStart");
}

@Override
public void onRestart() {
    super.onRestart();
    Log.d("lifecycle", "onRestart");
}

@Override
public void onResume() {
    super.onResume();
    Log.d("lifecycle", "onResume");
}

@Override
public void onPause() {
    super.onPause();
    Log.d("lifecycle", "onPause");
}

@Override
public void onStop() {
    super.onStop();
    Log.d("lifecycle", "onStop");
}

@Override
public void onDestroy() {
    super.onDestroy();
    Log.d("lifecycle", "onDestroy");
}
}
```

Note that if you override an activity's lifecycle method, you must call the overridden method in the parent class.

Before you run this application, create a Logcat message filter to show only messages from the application, filtering out system messages, by following these steps.

1. Select Debug from the Log level drop-down list.
2. Type in the filter text, in this case "lifecycle," in the search box. Figure 2.2 shows the **Logcat** window.

Figure 2.2: Creating a Logcat message filter

Run the application and notice the orientation of the application. It should be landscape. Now, try running another application and then switch back to the ActivityDemo application. Check the messages printed in Logcat.

Note that when you create a new application using Android Studio, the activity class may not extend **Activity** but **ActionBarActivity**. **ActionBarActivity** is a class in the Support Library that supports using the action bar in pre-3.0 Android devices. (The action bar is discussed in Chapter 6, "The Action Bar.") If you are not using the action bar or do not plan on deploying to pre-3.0 Android devices, you can replace **ActionBarActivity** with **Activity**.

Changing the Application Icon

If you do not like the application icon you have chosen, you can easily change it by following these steps.

- Save a jpeg or png file in **res/drawable**. Png is preferred because the format supports transparency.
- Edit the **android:icon** attribute of the manifest to point to the new image. You can refer to the image file with this format: **@drawable/***fileName*, where *fileName* is the name of the image file without the extension.

Using Android Resources

Android is rich, it comes with a bunch of assets (resources) you can use in your apps. To browse the available resources, open the application manifest in Android Studio and fill a property value by typing "**@android:**" followed by Ctrl+space. Android Studio will show the list of assets. (See Figure 2.3).

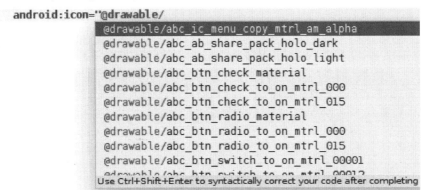

Figure 2.3: Using Android assets

For example, to see what images/icons are available, select **@android:drawable/**. To use a different icon for an application, change the value of the **android:icon** attribute.

```
android:icon="@android:drawable/ic_menu_day"
```

Starting Another Activity

The main activity of an Android application is started by the system itself, when the user selects the app icon from the Home screen. In an application with multiple activities, it is possible (and easy) to start another activity. In fact, starting an activity from another activity can be done simply by calling the **startActivity** method like this.

```
startActivity(intent);
```

where *intent* is an instance of the **android.content.Intent** class.

As an example, consider the SecondActivityDemo project that accompanies this book. It has two activities, **MainActivity** and **SecondActivity**. **MainActivity** contains a button that when clicked starts **SecondActivity**. This project also shows how you can write an event listener programmatically.

The manifest for **SecondActivityDemo** is given in Listing 2.3.

Listing 2.3: The manifest for SecondActivityDemo

```
<?xml version="1.0" encoding="utf-8"?>
<manifest xmlns:android="http://schemas.android.com/apk/res/android"
    package="com.example.secondactivitydemo"
    android:versionCode="1"
    android:versionName="1.0" >

    <uses-sdk
        android:minSdkVersion="8"
        android:targetSdkVersion="19" />

    <application
        android:allowBackup="true"
        android:icon="@drawable/ic_launcher"
        android:label="@string/app_name"
        android:theme="@style/AppTheme" >
        <activity
            android:name="com.example.secondactivitydemo.MainActivity"
            android:label="@string/app_name" >
            <intent-filter>
                <action android:name="android.intent.action.MAIN"/>
                <category
    android:name="android.intent.category.LAUNCHER"/>
            </intent-filter>
        </activity>
        <activity
            android:name="com.example.secondactivitydemo.SecondActivity"
            android:label="@string/title_activity_second" >
        </activity>
```

```
    </application>
</manifest>
```

Unlike the previous application, this project has two activities, one of which is declared as the main activity.

The layout files for the main and second activities are listed in Listings 2.4 and 2.5, respectively.

Listing 2.4: The activity_main.xml file

```
<RelativeLayout
    xmlns:android="http://schemas.android.com/apk/res/android"
    xmlns:tools="http://schemas.android.com/tools"
    android:layout_width="match_parent"
    android:layout_height="match_parent"
    android:paddingBottom="@dimen/activity_vertical_margin"
    android:paddingLeft="@dimen/activity_horizontal_margin"
    android:paddingRight="@dimen/activity_horizontal_margin"
    android:paddingTop="@dimen/activity_vertical_margin"
    tools:context=".MainActivity" >

    <TextView
        android:id="@+id/textView1"
        android:layout_width="wrap_content"
        android:layout_height="wrap_content"
        android:text="@string/first_screen" />

</RelativeLayout>
```

Listing 2.5: The activity_second.xml file

```
<RelativeLayout
        xmlns:android="http://schemas.android.com/apk/res/android"
    xmlns:tools="http://schemas.android.com/tools"
    android:layout_width="match_parent"
    android:layout_height="match_parent"
    android:paddingBottom="@dimen/activity_vertical_margin"
    android:paddingLeft="@dimen/activity_horizontal_margin"
    android:paddingRight="@dimen/activity_horizontal_margin"
    android:paddingTop="@dimen/activity_vertical_margin"
    tools:context=".SecondActivity" >

    <TextView
        android:id="@+id/textView1"
        android:layout_width="wrap_content"
        android:layout_height="wrap_content" />

</RelativeLayout>
```

Both activities contain a **TextView**. Touching the **TextView** in the main activity starts the second activity and pass a message to the latter. The second activity displays the message in its **TextView**.

The activity class for the main activity is given in Listing 2.6.

Listing 2.6: The MainActivity class

```
package com.example.secondactivitydemo;
import android.app.Activity;
import android.content.Intent;
import android.os.Bundle;
import android.view.Menu;
import android.view.MotionEvent;
import android.view.View;
import android.view.View.OnTouchListener;
import android.widget.TextView;

public class MainActivity extends Activity implements
        OnTouchListener {

    @Override
    protected void onCreate(Bundle savedInstanceState) {
        super.onCreate(savedInstanceState);
        setContentView(R.layout.activity_main);
        TextView tv = (TextView) findViewById(R.id.textView1);
        tv.setOnTouchListener(this);
    }

    @Override
    public boolean onCreateOptionsMenu(Menu menu) {
        // Inflate the menu; this adds items to the action bar if it
        // is present.
        getMenuInflater().inflate(R.menu.menu_main, menu);
        return true;
    }

    @Override
    public boolean onTouch(View arg0, MotionEvent event) {
        Intent intent = new Intent(this, SecondActivity.class);
        intent.putExtra("message", "Message from First Screen");
        startActivity(intent);
        return true;
    }
}
```

To handle the touch event, the **MainActivity** class has implemented the **OnTouchListener** interface and overridden its **onTouch** method. In this method, you create an **Intent** and put a message in it. You then call the **startActivity** method to start the second activity.

The **SecondActivity** class is given in Listing 2.7.

Listing 2.7: The SecondActivity class

```
package com.example.secondactivitydemo;
import android.app.Activity;
import android.content.Intent;
import android.os.Bundle;
import android.view.Menu;
import android.widget.TextView;
```

```
public class SecondActivity extends Activity {

    @Override
    protected void onCreate(Bundle savedInstanceState) {
        super.onCreate(savedInstanceState);
        setContentView(R.layout.activity_second);
        Intent intent = getIntent();
        String message = intent.getStringExtra("message");
        ((TextView) findViewById(R.id.textView1)).setText(message);
    }

    @Override
    public boolean onCreateOptionsMenu(Menu menu) {
        getMenuInflater().inflate(R.menu.menu_second, menu);
        return true;
    }
}
```

In the **onCreate** method of **SecondActivity**, you set the view content as usual. You then call the **getIntent** method and retrieve a message from its **getStringExtra** method, which you then pass to the **setText** method of the **TextView**. You retrieve the **TextView** by calling the **findViewById** method.

The main activity and the second activity are shown in Figures 2.4 and 2.5, respectively.

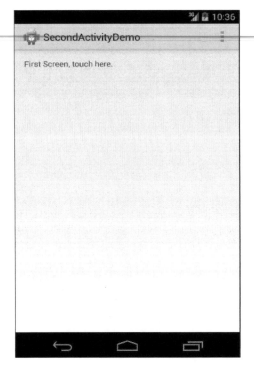

Figure 2.4: The main activity in SecondActivityDemo

Figure 2.5: The second activity in SecondActivityDemo

Activity-Related Intents

In the SecondActivityDemo project, you learned that you can start a new activity by passing an intent to the **startActivity** method. You can also call **startActivityForResult** if you want a result from the invoked activity.

Here is the code that activates an activity in the project:

```
Intent intent = new Intent(this, SecondActivity.class);
startActivity(intent);
```

And often you want to pass additional information to the invoked activity, which you can do by attaching the information to the intent. In the previous example, you did so by calling the **putExtra** method on the **Intent**:

```
Intent intent = new Intent(this, SecondActivity.class);
intent.putExtra("message", "Message from First Screen");
startActivity(intent);
```

An intent that is constructed by passing to it an activity class is called an explicit intent. The **Intent** in SecondActivityDemo is such an example.

You can also create an implicit intent, in which case you do not specify an activity class. Rather, you pass to the **Intent** class's constructor an action, such as **ACTION_SEND**, and let the system decide which activity to start. If there is more than one activities that can handle the intent, the system will normally ask the user to choose.

ACTION_SEND is a constant in the **Intent** class. Table 2.1 shows a list of actions that can start an activity as defined in the **Intent** class.

Action	Description
ACTION_MAIN	Start the activity as a main entry point.
ACTION_VIEW	View the data attached to the intent.
ACTION_ATTACH_DATA	Attach the data that has been added to the intent to some other place.
ACTION_EDIT	Edit the data attached to the intent.
ACTION_PICK	Pick an item from the data.
ACTION_CHOOSER	Displays all applications that can handle the intent.
ACTION_GET_CONTENT	Allow the user to select a particular kind of data and return it.
ACTION_DIAL	Dial the number attached to the intent.
ACTION_CALL	Call the person specified in the intent.
ACTION_SEND	Send the data attached to the intent.
ACTION_SENDTO	Send a message to the person specified in the intent data.
ACTION_ANSWER	Answer the incoming call.
ACTION_INSERT	Insert an empty item into the specified container.
ACTION_DELETE	Delete the specified data from its container.
ACTION_RUN	Run the attached data.
ACTION_SYNC	Perform a data synchronization.
ACTION_PICK_ACTIVITY	Select an activity from a set of activities.
ACTION_SEARCH	Perform a search using the specified string as the search key.
ACTION_WEB_SEARCH	Perform a web search using the specified string as the search key.
ACTION_FACTORY_TEST	Indicate this is the main entry point for factory tests.

Table 2.1: Intent actions for starting an activity

Not all intents can be used to start an activity. To make sure an **Intent** can revolve to an activity, call its **resolveActivity** method before passing it to **startActivity**:

```
if (intent.resolveActivity(getPackageManager()) != null) {
    startActivity(intent);
}
```

An intent that cannot resolve to an action will throw an exception if passed to **startActivity**.

For example, here is an Intent to send an email.

```
Intent intent = new Intent(Intent.ACTION_SEND);
intent.setType("message/rfc822"); // required
intent.putExtra(Intent.EXTRA_EMAIL,
        new String[] {"walter@example.com"}); // optional
intent.putExtra(Intent.EXTRA_SUBJECT, "subject"); // optional
intent.putExtra(Intent.EXTRA_TEXT   , "body"); // optional
```

```
// Verify that the intent will resolve to an activity
if (intent.resolveActivity(getPackageManager()) != null) {
    startActivity(intent);
} else {
    Toast.makeText(this, "No email client found.",
            Toast.LENGTH_LONG).show();
}
```

If multiple applications can handle an **Intent**, the user will be able to decide whether to always use the selected application in the future or to use it just for this occasion. You can force a chooser to appear each time (regardless of whether or not the user has decided to use the same app), by using this code:

```
startActivity(Intent.createChooser(intent, dialogTitle));
```

where *dialogTitle* is the title of the Chooser dialog.

As another example, the following code sends an ACTION_WEB_SEARCH intent. Upon receiving the message, the system will open the default web browser and tell the browser to google the search key.

```
String searchKey = "Buffalo";
Intent intent = new Intent(Intent.ACTION_WEB_SEARCH );
intent.putExtra(SearchManager.QUERY, searchKey);
startActivity(intent);
```

Summary

In this chapter you learned about the activity lifecycle and created two applications. The first application allowed you to observe when each of the lifecycle methods was called. The second application showed how to start an activity from another activity.

Chapter 3
UI Components

One of the first things you do when creating an Android application is build the user interface (UI) for the main activity. This is a relatively easy task thanks to the ready-to-use UI components that Android provides.

This chapter discusses some of the more important UI components.

Overview

The Android SDK provides various UI components called widgets that include many simple and complex components. Examples of widgets include buttons, text fields, progress bars, etc. In addition, you also need to choose a layout for laying out your UI components. Both widgets and layouts are implementations of the **android.view.View** class. A view is a rectangular area that occupies the screen. **View** is one of the most important Android types. However, unless you are creating a custom view, you don't often work with this class directly. Instead, you often spend time choosing and using layouts and UI components for your activities.

Figure 3.1 shows some Android UI components.

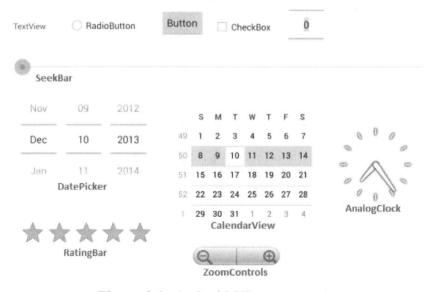

Figure 3.1: Android UI components

Using the Android Studio UI Tool

Creating a UI is easy with Android Studio. All you need is open the layout file for an activity and drag and drop UI components to the layout. The activity files are located in the **res/layout** directory of your application.

Figure 3.2 shows the UI tool for creating Android UI. This is what you see when you open an activity file. The tool window is divided into three main areas. On the left are the widgets, which are grouped into different categories such as Layouts, Widgets, Text Fields, Containers, etc. Click on the tab header of a category to see what widgets are available for that category.

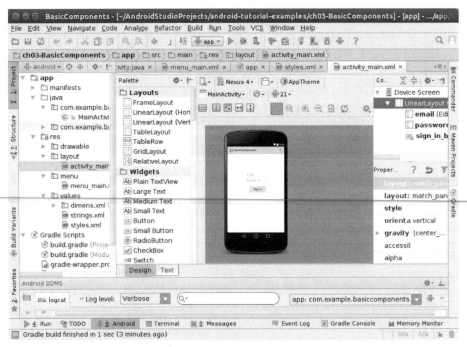

Figure 3.2: Using the UI tool

To choose a widget, click on the widget and drag it to the activity screen at the center. The screen in Figure 3.2 shows two text fields and a button. You can also view how your screen will look like in different devices by choosing a device from the **Devices** drop-down.

Each widget and layout has a set of properties derived from the **View** class or added to the implementation class. To change any of these properties, click on the widget on the drawing area or select it from the Outline pane in the Structure window on the right. The properties are listed in the small pane under the Layout pane.

What you do with the UI tool is reflected in the layout file, in the form of XML elements. To see what has been generated for you, click the XML view at the bottom of the UI tool.

Using Basic Components

The BasicComponents project is a simple Android application with one activity. The activity screen contains two text fields and a button.

You can either open the accompanying application or create one yourself by following the instructions in Chapter 1, "Getting Started." I will explain this project by presenting the manifest for the application, which is an XML file named **AndroidManifest.xml** located directly under the root directory.

Listing 3.1 shows the **AndroidManifest.xml** for the BasicComponents project.

Listing 3.1: The manifest for BasicComponents

```xml
<?xml version="1.0" encoding="utf-8"?>
<manifest xmlns:android="http://schemas.android.com/apk/res/android"
    package="com.example.basiccomponents"
    android:versionCode="1"
    android:versionName="1.0">

    <uses-sdk
        android:minSdkVersion="8"
        android:targetSdkVersion="17" />

    <application
        android:allowBackup="true"
        android:icon="@drawable/ic_launcher"
        android:label="@string/app_name"
        android:theme="@style/AppTheme" >
        <activity
            android:name="com.example.basiccomponents.MainActivity"
            android:label="@string/app_name" >
            <intent-filter>
                <action android:name="android.intent.action.MAIN"/>
                <category
                    android:name="android.intent.category.LAUNCHER"/>
            </intent-filter>
        </activity>
    </application>
</manifest>
```

The first thing to note is the **package** attribute of the **manifest** tag, which specifies **com.example.basiccomponents** as the Java package for the generated classes. Also note that the **application** element defines one activity, the main activity. The **application** element also specifies the icon, label, and theme for this application.

```
android:icon="@drawable/ic_launcher"
android:label="@string/app_name"
android:theme="@style/AppTheme">
```

It is good practice to reference a resource (such as an icon or a label) indirectly, like what I am doing here. **@drawable/ic_launcher**, the value for **android:icon**, refers to a drawable (normally an image file) that resides under the **res/drawable** directory. **ic_launcher** can mean an **ic_launcher.png** or **ic_launcher.jpg** file.

All string references start with **@string**. In the example above, **@string/app_name** refers to the **app_name** key in the **res/values/strings.xml** file. For this application, the **strings.xml** file is given in Listing 3.2.

Listing 3.2: The strings.xml file under res/values

```xml
<?xml version="1.0" encoding="utf-8"?>
<resources>
    <string name="app_name">BasicComponents</string>
    <string name="action_settings">Settings</string>
    <string name="prompt_email">Email</string>
    <string name="prompt_password">Password</string>
    <string name="action_sign_in"><b>Sign in</b></string>
</resources>
```

Let's now look at the main activity. There are two resources concerned with an activity, the layout file for the activity and the Java class that derives from **android.app.Activity**. For this project, the layout file is given in Listing 3.3 and the activity class (**MainActivity**) in Listing 3.4.

Listing 3.3: The layout file

```xml
<LinearLayout
    xmlns:android="http://schemas.android.com/apk/res/android"
    xmlns:tools="http://schemas.android.com/tools"
    android:layout_width="match_parent"
    android:layout_height="match_parent"
    android:layout_gravity="center"
    android:gravity="center_horizontal"
    android:orientation="vertical"
    android:padding="120dp"
    tools:context=".MainActivity" >

    <EditText
        android:id="@+id/email"
        android:layout_width="match_parent"
        android:layout_height="wrap_content"
        android:hint="@string/prompt_email"
        android:inputType="textEmailAddress"
        android:maxLines="1"
        android:singleLine="true" />

    <EditText
        android:id="@+id/password"
        android:layout_width="match_parent"
        android:layout_height="wrap_content"
        android:hint="@string/prompt_password"
        android:imeActionId="@+id/login"
        android:imeOptions="actionUnspecified"
        android:inputType="textPassword"
        android:maxLines="1"
        android:singleLine="true" />
```

```
<Button
    android:id="@+id/sign_in_button"
    android:layout_width="wrap_content"
    android:layout_height="wrap_content"
    android:layout_gravity="right"
    android:layout_marginTop="16dp"
    android:paddingLeft="32dp"
    android:paddingRight="32dp"
    android:text="@string/action_sign_in" />

</LinearLayout>
```

The layout file contains a **LinearLayout** element with three children, namely two
EditText components and a button.

Listing 3.4: The MainActivity class of Basic Components

```
package com.example.basiccomponents;
import android.os.Bundle;
import android.app.Activity;
import android.view.Menu;

public class MainActivity extends Activity {

    @Override
    protected void onCreate(Bundle savedInstanceState) {
        super.onCreate(savedInstanceState);
        setContentView(R.layout.activity_main);
    }

    @Override
    public boolean onCreateOptionsMenu(Menu menu) {
        // Inflate the menu; this adds items to the action bar if it
        // is present.
        getMenuInflater().inflate(R.menu.menu_main, menu);
        return true;
    }
}
```

The **MainActivity** class in Listing 3.4 is a boilerplate class created by Android Studio. It
overrides the **onCreate** and **onCreateOptionsMenu** methods. **onCreate** is a lifecycle
method that gets called when the application is created. In Listing 3.4, it simply sets the
content view for the activity using the layout file. **onCreateOptionsMenu** initializes the
content of the activity's options menu. It must return true for the menu to be displayed.

Run the application and you'll see the activity as shown in Figure 3.3.

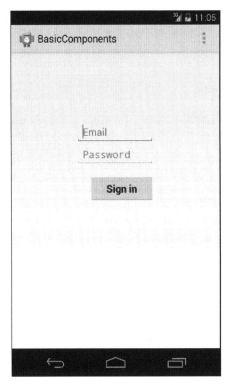

Figure 3.3: The BasicComponents project

Toast

A **Toast** is a small popup for displaying a message as feedback for the user. A **Toast** does not replace the current activity and only occupies the space taken by the message.

Figure 3.4 shows a Toast that says "Downloading file..." The **Toast** disappears after a prescribed period of time.

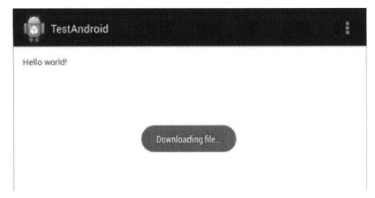

Figure 3.4: A Toast

The **android.widget.Toast** class is the template for creating a **Toast**. To create a **Toast**, call its only constructor that takes a **Context** as an argument:

```
public Toast(android.content.Context context)
```

Toast also provides two static **makeText** methods for creating an instance of **Toast**. The signatures of both method overloads are

```
public static Toast makeText (android.content.Context context,
        int resourceId, int duration)
```

```
public static Toast makeText (android.content.Context context,
        java.lang.CharSequence text, int duration)
```

Both overloads require that you pass a **Context**, possibly the active activity, as the first argument. In addition, both overloads take a string, which may come from a **strings.xml** file or a **String** object, and the duration of the display for the **Toast**. Two valid values for the duration are the **LENGTH_LONG** and **LENGTH_SHORT** static finals in **Toast**.

To display a **Toast**, call its **show** method. This method takes no argument.

The following code snippet shows how to create and display a **Toast** in an activity class.

```
Toast.makeText(this, "Downloading...", Toast.LENGTH_LONG).show();
```

By default, a **Toast** is displayed near the bottom of the active activity. However, you can change its position by calling its **setGravity** method before calling its **show** method. Here is the signature of **setGravity**.

```
public void setGravity(int gravity, int xOffset, int yOffset)
```

The valid value for *gravity* is one of the static finals in the **android.view.Gravity** class, including **CENTER_HORIZONTAL** and **CENTER_VERTICAL**.

You can also enhance the look of a **Toast** by creating your own layout file and passing it to the **Toast**'s **setView** method. Here is an example of how to create a custom **Toast**.

```
LayoutInflater inflater = getLayoutInflater();
View layout = inflater.inflate(R.layout.toast_layout,
        (ViewGroup) findViewById(R.id.toast_layout_root));
Toast toast = new Toast(getApplicationContext());
toast.setView(layout);
toast.show();
```

In this case, **R.layout.toast_layout** is the layout identifier for the Toast and **R.id.toast_layout_root** is the id of the **root** element in the layout file.

AlertDialog

Like a **Toast**, an **AlertDialog** is a window that provides feedback to the user. Unlike a **Toast** that fades by itself, however, an **AlertDialog** shows indefinitely until it loses focus. In addition, an **AlertDialog** can contain up to three buttons and a list of selectable items.

A button added to an **AlertDialog** can be connected to a listener that gets triggered when the button is clicked.

Figure 3.5 shows a sample **AlertDialog**.

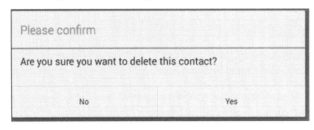

Figure 3.5: An AlertDialog

The **android.app.AlertDialog** class is the template for creating an **AlertDialog**. All constructors in this class are protected, so you cannot use them unless you are subclassing the class. Instead, you should use the **AlertDialog.Builder** class to create an **AlertDialog**. You can use one of the two constructors of **AlertDialog.Builder**.

```
public AlertDialog.Builder(android.content.Context context)
```

```
public AlertDialog.Builder(android.content.Context context,
        int theme)
```

Once you have an instance of **AlertDialog.Builder**, you can call its **create** method to return an **AlertDialog**. However, before calling **create** you can call various methods of **AlertDialog.Builder** to decorate the resulting **AlertDialog**. Interestingly, the methods in **AlertDialog.Builder** return the same instance of **AlertDialog.Builder**, so you can cascade them. Here are some of the methods in **AlertDialog.Builder**.

```
public AlertDialog.Builder setIcon(int resourceId)
```
> Sets the icon of the resulting **AlertDialog** with the Drawable pointed by *resourceId*.

```
public AlertDialog.Builder setMessage(java.lang.CharSequence message)
```
> Sets the message of the resulting **AlertDialog** .

```
public AlertDialog.Builder setTitle(java.lang.CharSequence title)
```
> Sets the title of the resulting **AlertDialog**.

```
public AlertDialog.Builder setNegativeButton(
        java.lang.CharSequence text,
        android.content.DialogInterface.OnClickListener listener)
```
> Assigns a button that the user should click to provide a negative response.

```
public AlertDialog.Builder setPositiveButton(
        java.lang.CharSequence text,
        android.content.DialogInterface.OnClickListener listener)
```
> Assigns a button that the user should click to provide a positive response.

```
public AlertDialog.Builder setNeutralButton(
        java.lang.CharSequence text,
        android.content.DialogInterface.OnClickListener listener)
```
> Assigns a button that the user should click to provide a neutral response.

For instance, the following code produces an **AlertDialog** that looks like the one in Figure

3.5.

```
new AlertDialog.Builder(this)
    .setTitle("Please confirm")
    .setMessage(
        "Are you sure you want to delete " +
        "this contact?")
    .setPositiveButton("Yes",
        new DialogInterface.OnClickListener() {
            public void onClick(
                    DialogInterface dialog,
                    int whichButton) {

                // delete picture here

                dialog.dismiss();
            }
        })
    .setNegativeButton("No",
        new DialogInterface.OnClickListener() {
            public void onClick(
                    DialogInterface dialog,
                    int which) {
                dialog.dismiss();
            }
        })
    .create()
    .show();
```

Pressing the Yes button will execute the listener passed to the **setPositiveButton** method and pressing the No button will run the listener passed to the **setNegativeButton** method.

Notifications

A notification is a message on the status bar. Unlike a toast, a notification is persistent and will keep showing until it is closed or the device is shut down.

A notification is an instance of **android.app.Notification**. The most convenient way to create a notification is by using a nested class called **Builder**, which can be instantiated by passing a **Context**. You can then call the **build** method on the builder to create a **Notification**.

```
Notification n = new Notification.Builder(context).build();
```

The **Notification.Builder** class has methods to decorate the resulting notification. These methods include **addAction**, **setAutoCancel**, **setColor**, **setContent**, **setContentTitle**, **setContentIntent**, **setLargeIcon**, **setSmallIcon** and **setSound**.

Many of these methods are self-explanatory, but **addAction** and **setContentIntent** are of particular importance because you can use them to add an action that will be performed when the user touches the notification. In this case, a notification action is represented by a **PendingIntent**. Here are the signatures of **addAction** and

setContentIntent, both of which take a **PendingIntent**.

```
public Notification.Builder addAction(int icon,
        java.lang.CharSequence title,
        android.app.PendingIntent intent)

public Notification.Builder setContentIntent(
        android.app.PendingIntent intent)
```

When the user touches the notification, the **send** method of the **PendingIntent** will be invoked. See the sidebar for a description of **PendingIntent**.

setAutoCancel is also important and passing **true** to it allows the notification to be dismissed when the user touches it on the notification drawer. The notification drawer is an area that opens when you slide down the status bar. The notification drawer shows all notifications that the system have received and have not been dismissed.

The methods in **Notification.Builder** return the same **Builder** object, so they can be cascaded:

```
Notification notification  = new Notification.Builder(context)
        .setContentTitle("New notification")
        .setContentText("You've got one!")
        .setSmallIcon(android.R.drawable.star_on)
        .setContentIntent(pendingIntent)
        .setAutoCancel(false)
        .addAction(android.R.drawable.star_big_on,
                "Open", pendingIntent)
        .build();
```

To sound a ringtone, flash lights and make the device vibrate, you can OR the defaults flags like so:

```
notification.defaults|= Notification.DEFAULT_SOUND;
notification.defaults|= Notification.DEFAULT_LIGHTS;
notification.defaults|= Notification.DEFAULT_VIBRATE;
```

In addition, to make repeating sound, you can set the FLAG_INSISTENT flag.

```
notification.flags |= Notification.FLAG_INSISTENT;
```

To publish a notification, use the **NotificationManager**, which is one of the built-in services in the Android system. As it is an existing system service, you can obtain it by calling the **getSystemService** method on an activity, like so:

```
NotificationManager notificationManager = (NotificationManager)
        getSystemService(NOTIFICATION_SERVICE);
```

Then, you can publish a notification by calling the **notify** method on the **NotificationManager**, passing a unique ID and the notification.

```
notificationManager.notify(notificationId, notification);
```

The PendingIntent Class

A **PendingIntent** encapsulates an **Intent** and an action that will be carried out when its **send** method is invoked. Since a **PendingIntent** is a pending intent, the action is normally an operation that will be invoked some time in the future, most probably by the system. For example, a **PendingIntent** can be used to construct a **Notification** so that something can be made happen when the user touches the notification.

The action in a **PendingIntent** is one of several methods in the **Context** class, such as **startActivity**, **startService** or **sendBroadcast**.

You have learned that to start an activity you can pass an **Intent** to the **startActivity** method on a **Context**.

```
Intent intent = ...
context.startActivity(intent);
```

The equivalent code for starting an activity using a **PendingIntent** looks like this:

```
Intent intent = ...
PendingIntent pendingIntent = PendingIntent.getActivity(context, 0,
intent, 0);
pendingIntent.send();
```

The static method **getActivity** is one of several methods that returns an instance of **PendingIntent**. Other methods are **getActivities**, **getService** and **getBroadcast**. These methods determine the action that the resulting **PendingIntent** can perform. Constructing a **PendingIntent** by calling **getActivity** returns an instance that can start an activity. Creating a **PendingIntent** using **getService** gives you an instance that can be used to start a service. You call **getBroadcast** if you want a **PendingIntent** for sending a broadcast.

The notification ID is an integer that you can choose. This ID is needed just in case you want to cancel the notification, in which case you pass the ID to the **cancel** method of the **NotificationManager**:

```
notificationManager.cancel(notificationId);
```

The NotificationDemo project is an application that shows how to use notifications. The main activity of the app contains two buttons, one for publishing a notification and one for cancelling it. After the notification is published, opening it will invoke a second activity.

Listing 3.5 shows the layout file and Listing 3.6 the activity class.

Listing 3.5: The layout file of the main activity of NotificationDemo

```
<LinearLayout
    xmlns:android="http://schemas.android.com/apk/res/android"
    xmlns:tools="http://schemas.android.com/tools"
    android:layout_width="wrap_content"
    android:layout_height="wrap_content"
    android:orientation="horizontal">

    <Button
        android:layout_width="wrap_content"
        android:layout_height="wrap_content"
        android:onClick="setNotification"
        android:text="Set Notification" />
```

```
    <Button
        android:layout_width="wrap_content"
        android:layout_height="wrap_content"
        android:onClick="clearNotification"
        android:text="Clear Notification" />
</LinearLayout>
```

Listing 3.6: The main activity class

```java
package com.example.notificationdemo;
import android.app.Activity;
import android.app.Notification;
import android.app.NotificationManager;
import android.app.PendingIntent;
import android.content.Intent;
import android.os.Bundle;
import android.view.Menu;
import android.view.View;

public class MainActivity extends Activity {
    int notificationId = 1001;

    @Override
    protected void onCreate(Bundle savedInstanceState) {
        super.onCreate(savedInstanceState);
        setContentView(R.layout.activity_main);
    }

    @Override
    public boolean onCreateOptionsMenu(Menu menu) {
        getMenuInflater().inflate(R.menu.menu_main, menu);
        return true;
    }

    public void setNotification(View view) {
        Intent intent = new Intent(this, SecondActivity.class);
        PendingIntent pendingIntent =
                PendingIntent.getActivity(this, 0, intent, 0);

        Notification notification  = new Notification.Builder(this)
                .setContentTitle("New notification")
                .setContentText("You've got a notification!")
                .setSmallIcon(android.R.drawable.star_on)
                .setContentIntent(pendingIntent)
                .setAutoCancel(true)
                .addAction(android.R.drawable.ic_menu_gallery,
                        "Open", pendingIntent)
                .build();
        NotificationManager notificationManager =
                (NotificationManager) getSystemService(
                        NOTIFICATION_SERVICE);
        notificationManager.notify(notificationId, notification);
    }
```

```
public void clearNotification(View view) {
    NotificationManager notificationManager =
            (NotificationManager) getSystemService(
                    NOTIFICATION_SERVICE);
    notificationManager.cancel(notificationId);
}
}
```

When you run the application, you will see the main activity with two buttons, like the one shown in Figure 3.6.

Figure 3.6: The NotificationDemo project

If you click the Set Notification button, the notification icon (an orange star) will be shown on the status bar.

Figure 3.6: The notification icon on the status bar

Now drag down the status bar right to the bottom of the screen to open the notification drawer (See Figure 3.7).

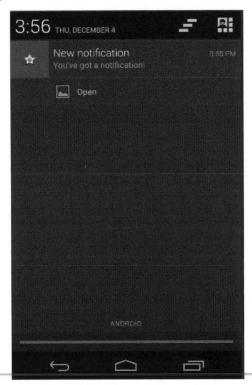

Figure 3.7: The notification in the notification drawer

The notification drawer in Figure 3.7 has one notification. The upper part of the notification UI has a title that screams "New notification." The text "You've got a notification" is the notification content. The lower part is a UI component that represents an action. With its **autoClose** set to true, the notification will be canceled if the user touches on the content. However, it will not close if the user touches on the action UI. Unfortunately, it is not straightforward to make both areas cancel the notification drawer when touched. To remedy this situation you can use a broadcast receiver, as explained in Chapter 25, "Broadcast Receivers."

In this example, touching on the notification starts the SecondActivity, as shown in Figure 3.8.

Figure 3.8: The second activity activated by the notification

Summary

In this chapter you learned about the UI components available in Android. You also learned how to use the toast, dialogs and notifications.

Chapter 4
Layouts

Layouts are important because they directly affect the look and feel of your application. Technically, a layout is a view that arranges child views added to it. Android comes with a number of built-in layouts, ranging from **LinearLayout**, which is the easiest to use, to **RelativeLayout**, which is the most powerful.

This chapter discusses the various layouts in Android.

Overview

An important Android component, a layout defines the visual structure of your UI components. A layout is a subclass of **android.view.ViewGroup**, which in turn derives from **android.view.View**. A **ViewGroup** is a special view that can contain other views. A layout can be declared in a layout file or added programmatically at runtime.

The following are some of the layouts in Android.

- **LinearLayout**. A layout that aligns its children in the same direction, either horizontally or vertically.
- **RelativeLayout**. A layout that arranges each of its children based on the positions of one or more of its siblings.
- **FrameLayout**. A layout that arranges each of its children based on top of one another.
- **TableLayout**. A layout that organizes its children into rows and columns.
- **GridLayout**. A layout that arranges its children in a grid.

In a majority of cases, a view in a layout must have the **layout_width** and **layout_height** attributes so that the layout knows how to size the view. Both **layout_width** and **layout_height** attributes may be assigned the value **match_parent** (to match the parent's width/height), **wrap_content** (to match the width/height of its content) or a measurement unit.

The **AbsoluteLayout**, which offers exact locations for its child views, is deprecated and should not be used. Use **RelativeLayout** instead.

LinearLayout

A **LinearLayout** is a layout that arranges its children either horizontally or vertically, depending on the value of its **orientation** property. The **LinearLayout** is the easiest layout to use.

The layout in Listing 4.1 is an example of **LinearLayout** with horizontal orientation. It contains three children, an **ImageButton**, a **TextView** and a **Button**.

Listing 4.1: A horizontal LinearLayout

```
<LinearLayout
    xmlns:android="http://schemas.android.com/apk/res/android"
    xmlns:tools="http://schemas.android.com/tools"
    android:orientation="horizontal"
    android:layout_width="match_parent"
    android:layout_height="match_parent">

    <ImageButton
        android:src="@android:drawable/btn_star_big_on"
        android:layout_width="wrap_content"
        android:layout_height="wrap_content"/>

    <TextView
        android:layout_width="wrap_content"
        android:layout_height="wrap_content"
        android:text="@string/hello_world" />
    <Button android:text="Button1"
        android:layout_width="wrap_content"
        android:layout_height="wrap_content"/>

</LinearLayout>
```

Figure 4.1 shows the views in the **LinearLayout** in Listing 4.1.

The layout in Listing 4.2 is a vertical **LinearLayout** with three child views, an **ImageButton**, a **TextView** and a **Button**.

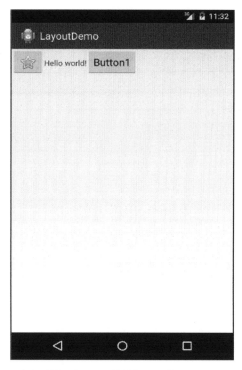

Figure 4.1: Horizontal LinearLayout example

Listing 4.2: Vertical linear layout

```
<LinearLayout
    xmlns:android="http://schemas.android.com/apk/res/android"
    xmlns:tools="http://schemas.android.com/tools"
    android:orientation="vertical"
    android:layout_width="match_parent"
    android:layout_height="match_parent">

    <ImageButton
        android:src="@android:drawable/btn_star_big_on"
        android:layout_gravity="center"
        android:layout_width="wrap_content"
        android:layout_height="wrap_content"/>
    <TextView
        android:layout_gravity="center"
        android:layout_width="wrap_content"
        android:layout_height="wrap_content"
        android:layout_marginLeft="15dp"
        android:text="@string/hello_world"/>
    <Button android:text="Button1"
        android:layout_gravity="center"
        android:layout_width="wrap_content"
        android:layout_height="wrap_content"/>
</LinearLayout>
```

Figure 4.2 shows the vertical **LinearLayout**.

Figure 4.2: Vertical linear layout example

Note that each view in a layout can have a **layout_gravity** attribute to determine its position within its axis. For example, setting the **layout_gravity** attribute to **center** will center it.

A **LinearLayout** can also have a **gravity** attribute that affects its gravity. For example, the layout in Listing 4.3 is a vertical **LinearLayout** whose **gravity** attribute is set to bottom.

Listing 4.3: Vertical linear layout with bottom gravity

```
<LinearLayout
    xmlns:android="http://schemas.android.com/apk/res/android"
    xmlns:tools="http://schemas.android.com/tools"
    android:orientation="vertical"
    android:layout_width="match_parent"
    android:layout_height="match_parent"
    android:gravity="bottom">

    <ImageButton
        android:src="@android:drawable/btn_star_big_on"
        android:layout_gravity="center"
        android:layout_width="wrap_content"
        android:layout_height="wrap_content"/>
    <TextView
        android:layout_gravity="center"
        android:layout_width="wrap_content"
        android:layout_height="wrap_content"
        android:layout_marginLeft="15dp"
```

```
        android:text="@string/hello_world"/>
    <Button android:text="Button1"
        android:layout_gravity="center"
        android:layout_width="wrap_content"
        android:layout_height="wrap_content"/>
</LinearLayout>
```

Figure 4.3 shows the vertical **LinearLayout** in Listing 4.3.

Figure 4.3: Vertical linear layout with gravity

RelativeLayout

The **RelativeLayout** is the most powerful layout available. All children in a **RelativeLayout** can be positioned relative to each other or to their parent. For example, you can tell a view to be positioned to the left or right of another view. Or, you can specify that a view is aligned to the bottom or top edge of its parent.

Positioning a child in a **RelativeLayout** is achieved using the attributes summarized in Table 4.1.

Attribute	Description
layout_above	Places the bottom edge of this view above the specified view ID.
layout_alignBaseline	Places the baseline of this view on the baseline of the specified view ID.
layout_alignBottom	Aligns the bottom of this view with the specified view.
layout_alignEnd	Aligns the end edge of this view with the end edge of the specified view.
layout_alignLeft	Aligns the left edge of this view with the left edge of the specified view.
layout_alignParentBottom	A value of true aligns the bottom of this view with the bottom edge of its parent.
layout_alignParentEnd	A value of true aligns the end edge of this view with the end edge of its parent.
layout_alignParentLeft	A value of true aligns the left edge of this view with the left edge of its parent.
layout_alignParentRight	A value of true aligns the right edge of this view with the right edge of its parent.
layout_alignParentStart	A value of true aligns the start edge of this view with the start edge of its parent.
layout_alignParentTop	A value of true aligns the top edge of this view with the top edge of its parent.
layout_alignRight	Aligns the right edge of this view with the right edge of the given view.
layout_alignStart	Aligns the start edge of this view with the start edge of the given view.
layout_alignTop	Aligns the top edge of this view with the top edge of the given view.
layout_alignWithParentIfMissing	A value of true sets the parent as the anchor when the anchor cannot be found for layout_toLeftOf, layout_toRightOf, etc.
layout_below	Places the top edge of this view below the given view.
layout_centerHorizontal	A value of true centers this view horizontally within its parent.
layout_centerInParent	A value of true centers this view horizontally and vertically within its parent.
layout_centerVertical	A value of true center this view vertically within its parent.
layout_toEndOf	Places the start edge of this view to the end of the given view.
layout_toLeftOf	Places the right edge of this view to the left of the given view.
layout_toRightOf	Places the left edge of this view to the right of the given view.
layout_toStartOf	Places the end edge of this view to the start of the given view.

Table 4.1: Attributes for children of a RelativeLayout

As an example, the layout in Listing 4.4 specifies the positions of three views and a **RelativeLayout**.

Listing 4.4: Relative layout

```
<RelativeLayout
        xmlns:android="http://schemas.android.com/apk/res/android"
    xmlns:tools="http://schemas.android.com/tools"
    android:layout_width="match_parent"
    android:layout_height="match_parent"
    android:paddingLeft="2dp"
    android:paddingRight="2dp">
```

```xml
    <Button
        android:id="@+id/cancelButton"
        android:layout_width="wrap_content"
        android:layout_height="wrap_content"
        android:text="Cancel" />

    <Button
        android:id="@+id/saveButton"
        android:layout_width="wrap_content"
        android:layout_height="wrap_content"
        android:layout_toRightOf="@id/cancelButton"
        android:text="Save" />

    <ImageView
        android:layout_width="150dp"
        android:layout_height="150dp"
        android:layout_marginTop="230dp"
        android:padding="4dp"
        android:layout_below="@id/cancelButton"
        android:layout_centerHorizontal="true"
        android:src="@android:drawable/ic_btn_speak_now" />

    <LinearLayout
        android:id="@+id/filter_button_container"
        android:layout_width="match_parent"
        android:layout_height="wrap_content"
        android:layout_alignParentBottom="true"
        android:gravity="center|bottom"
            android:background="@android:color/white"
        android:orientation="horizontal" >

        <Button
            android:id="@+id/filterButton"
            android:layout_width="wrap_content"
            android:layout_height="fill_parent"
            android:text="Filter" />

        <Button
            android:id="@+id/shareButton"
            android:layout_width="wrap_content"
            android:layout_height="fill_parent"
            android:text="Share" />

        <Button
            android:id="@+id/deleteButton"
            android:layout_width="wrap_content"
            android:layout_height="fill_parent"
            android:text="Delete" />
    </LinearLayout>
</RelativeLayout>
```

Adding An Identifier

The first button in Listing 4.4 includes the following **id** attribute so that it can be referenced from the code.

```
android:id="@+id/cancelButton"
```

The plus sign (+) after @ indicates that the identifier (in this case, **cancelButton**) is being added with this declaration and is not declared in a resource file.

Figure 4.4 shows the **RelativeLayout** in Listing 4.4.

Figure 4.4: RelativeLayout

FrameLayout

A **FrameLayout** positions its children on top of each other. By adjusting the margin and padding of a view, it is possible to lay out the view below another view, as shown by the layout in Listing 4.5.

Listing 4.5: Using a FrameLayout

```
<FrameLayout
    xmlns:android="http://schemas.android.com/apk/res/android"
    xmlns:tools="http://schemas.android.com/tools"
    android:orientation="horizontal"
```

```
        android:layout_width="match_parent"
        android:layout_height="match_parent">

    <Button android:text="Button1"
        android:layout_width="wrap_content"
        android:layout_height="wrap_content"
        android:layout_marginTop="100dp"
        android:layout_marginLeft="100dp" />
    <ImageButton
        android:src="@android:drawable/btn_star_big_on"
        android:alpha="0.35"
        android:layout_width="wrap_content"
        android:layout_height="wrap_content"
        android:layout_marginTop="90dp"
        android:layout_marginLeft="90dp" />
</FrameLayout>
```

The layout in Listing 4.5 uses a **FrameLayout** with a **Button** and an **ImageButton**. The **ImageButton** is placed above the **Button**, as shown in Figure 4.5.

Figure 4.5: Using FrameLayout

TableLayout

A **TableLayout** is used to arrange child views in rows and columns. The **TableLayout** class is a subclass of **LinearLayout**. To add a row in a **TableLayout**, use a **TableRow** element. A view directly added to a **TableLayout** (without a **TableRow**) will also occupy a row that spans all columns.

The layout in Listing 4.6 shows a **TableLayout** with four rows, two of which are created using **TableRow** elements.

Listing 4.6: Using the TableLayout

```
<TableLayout
        xmlns:android="http://schemas.android.com/apk/res/android"
    android:layout_width="wrap_content"
    android:layout_height="wrap_content"
    android:layout_gravity="center" >

    <TableRow
        android:id="@+id/tableRow1"
        android:layout_width="500dp"
        android:layout_height="wrap_content"
        android:padding="5dip" >

        <ImageView android:src="@drawable/ic_launcher" />
        <ImageView android:src="@android:drawable/btn_star_big_on" />
        <ImageView android:src="@drawable/ic_launcher" />
    </TableRow>

    <TableRow
        android:id="@+id/tableRow2"
        android:layout_width="wrap_content"
        android:layout_height="wrap_content" >

        <ImageView android:src="@android:drawable/btn_star_big_off" />
        <TextClock />
        <ImageView android:src="@android:drawable/btn_star_big_on" />
    </TableRow>

    <EditText android:hint="Your name" />

    <Button
        android:layout_height="wrap_content"
        android:text="Go" />

</TableLayout>
```

Figure 4.6 shows how the **TableLayout** in Listing 4.6 is rendered.

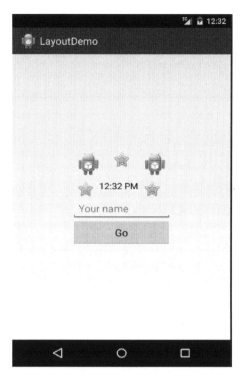

Figure 4.6: Using TableLayout

Grid Layout

A **GridLayout** is similar to a **TableLayout**, but the number of columns must be specified using a **columnCount** attribute. Listing 4.7 shows an example of **GridLayout**.

Listing 4.7: GridLayout example

```
<GridLayout
      xmlns:android="http://schemas.android.com/apk/res/android"
    android:layout_width="wrap_content"
    android:layout_height="wrap_content"
    android:layout_gravity="center"
    android:columnCount="3">

    <!-- 1st row, spanning 3 columns -->
    <TextView
        android:layout_width="wrap_content"
        android:layout_height="wrap_content"
        android:text="Enter your name"
        android:layout_columnSpan="3"
        android:textSize="26sp"
        />
    <!-- 2nd row -->
    <TextView android:text="First Name"/>
```

```
    <EditText
        android:id="@+id/firstName"
        android:layout_width="200dp"
        android:layout_columnSpan="2"/>

    <!-- 3rd row -->
    <TextView android:text="Last Name"/>
    <EditText
        android:id="@+id/lastName"
        android:layout_width="200dp"
        android:layout_columnSpan="2"/>

    <!-- 4th row, spanning 3 columns -->
    <Button
        android:layout_width="wrap_content"
        android:layout_height="wrap_content"
        android:layout_column="2"
        android:layout_gravity="right"
        android:text="Submit"/>
</GridLayout>
```

Figure 4.7 visualizes the **GridLayout** in Listing 4.7.

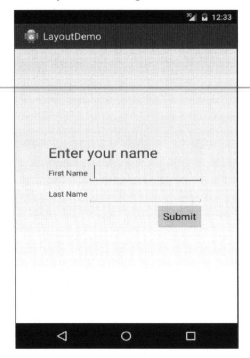

Figure 4.7: Using GridLayout

Creating A Layout Programmatically

The most common way to create a layout is by using an XML file, as you have seen in the examples in this chapter. However, it is also possible to create a layout programmatically, by instantiating the layout class and passing it to the **addContentView** method in an activity class. For instance, the following code is part of the **onCreate** method of an activity that programmatically creates a **LinearLayout**, sets a couple properties, and passes it to **addContentView**.

```
LinearLayout root = new LinearLayout(this);
LinearLayout.LayoutParams matchParent = new
        LinearLayout.LayoutParams(
                LinearLayout.LayoutParams.MATCH_PARENT,
                LinearLayout.LayoutParams.MATCH_PARENT);
root.setOrientation(LinearLayout.VERTICAL);
root.setGravity(Gravity.CENTER_VERTICAL);          addContentView(root,
matchParent);
```

Summary

A layout is responsible for arranging its child views. It directly affect the look and feel of an application. In this chapter you learned some of the layouts available in Android, **LinearLayout**, **RelativeLayout**, **FrameLayout**, **TableLayout**, and **GridLayout**.

Chapter 5
Listeners

Like many GUI systems, Android is event based. User interaction with a view in an activity may trigger an event and you can write code that gets executed when the event occurs. The class that contains code to respond to a certain event is called an event listener.

In this chapter you will learn how to handle events and write event listeners.

Overview

Most Android programs are interactive. The user can interact with the application easily thanks to the event-driven programming paradigm the Android framework offers. Examples of events that may occur when the user is interacting with an activity are click, long-click, touch, key, and so on.

To make a program respond to a certain event, you need to write a listener for that event. The way to do it is by implementing an interface that is nested in the **android.view.View** class. Table 5.1 shows some of the listener interfaces in **View** and the corresponding method in each interface that will get called when the corresponding event occurs.

Interface	Method
OnClickListener	onClick()
OnLongClickListner	OnLongClick()
OnFocusChangeListener	OnFocusChange()
OnKeyListener	OnKey()
OnTouchListener	OnTouch()

Table 5.1: Listener interfaces in View

Once you create an implementation of a listener interface, you can pass it to the appropriate **setOnXXXListener** method of the view you want to listen to, where *XXX* is the event name. For example, to create a click listener for a button, you would write this in your activity class.

```
private OnClickListener clickListener = new OnClickListener() {
    public void onClick(View view) {
        // code to execute in response to the click event
    }
};

protected void onCreate(Bundle savedValues) {
```

```
    ...
    Button button = (Button)findViewById(...);
    button.setOnClickListener(clickListener);
    ...
}
```

Alternatively, you can make your activity class implement the listener interface and provide an implementation of the needed method as part of the activity class.

```
public class MyActivity extends Activity
        implements View.OnClickListener {
    protected void onCreate(Bundle savedValues) {
        ...
        Button button = (Button)findViewById(...);
        button.setOnClickListener(this);
        ...
    }

    // implementation of View.OnClickListener
    @Override
    public void onClick(View view) {
        // code to execute in response to the click event
    }

    ...
}
```

In addition, there is a shortcut for handling the click event. You can use the **onClick** attribute in the declaration of the target view in the layout file and write a public method in the activity class. The public method must have no return value and take a **View** argument. For example, if you have this method in your activity class

```
public void showNote(View view) {
    // do something
}
```

you can use this **onClick** attribute in a view to attach the method to the click event of that view.

```
<Button android:onClick="showNote" .../>
```

In the background, Android will create an implementation of the **OnClickListener** interface and attach it to the view.

In the sample applications that follow you will learn how to write event listeners.

Note

A listener runs on the main thread. This means you should use a different thread if your listener takes a long time (say, more than 200ms) to run. Or else, your application will look unresponsive during the execution of the listener code. You have two choices for solving this. You can either use a handler or an **AsyncTask**. The handler is covered in Chapter 2, "Handling the Handler" and **AsyncTask** in Chapter 23, "Asynchronous Tasks." For long-running tasks, you should also consider using the Java Concurrency Utilities.

Using the onClick Attribute

As an example of using the **onClick** attribute to handle the click event of a view, consider the MulticolorClock project that accompanies this book. It is a simple application with a single activity that shows an analog clock that can be clicked to change its color. **AnalogClock** is one of the widgets available on Android, so writing the view for the application is a breeze. The main objective of this project is to demonstrate how to write a listener by using a callback method in the layout file.

The manifest for MulticolorClock is given in Listing 5.1. There is nothing out of ordinary here and you should not find it difficult to understand.

Listing 5.1: The manifest for MulticolorClock

```xml
<?xml version="1.0" encoding="utf-8"?>
<manifest xmlns:android="http://schemas.android.com/apk/res/android"
    package="com.example.multicolorclock"
    android:versionCode="1"
    android:versionName="1.0" >

    <uses-sdk
        android:minSdkVersion="8"
        android:targetSdkVersion="17" />

    <application
        android:allowBackup="true"
        android:icon="@drawable/ic_launcher"
        android:label="@string/app_name"
        android:theme="@style/AppTheme" >
        <activity
            android:name="com.example.multicolorclock.MainActivity"
            android:label="@string/app_name" >
            <intent-filter>
                <action android:name="android.intent.action.MAIN" />
                <category android:name="android.intent.category.LAUNCHER"
    />
            </intent-filter>
        </activity>
    </application>
</manifest>
```

Now comes the crucial part, the layout file. It is called **activity_main.xml** and located under the **res/layout** directory. The layout file is presented in Listing 5.2.

Listing 5.2: The layout file in MulticolorClock

```xml
<RelativeLayout
    xmlns:android="http://schemas.android.com/apk/res/android"
    xmlns:tools="http://schemas.android.com/tools"
    android:layout_width="match_parent"
    android:layout_height="match_parent"
    android:paddingBottom="@dimen/activity_vertical_margin"
```

```
    android:paddingLeft="@dimen/activity_horizontal_margin"
    android:paddingRight="@dimen/activity_horizontal_margin"
    android:paddingTop="@dimen/activity_vertical_margin"
    tools:context=".MainActivity">

    <AnalogClock
        android:id="@+id/analogClock1"
        android:layout_width="wrap_content"
        android:layout_height="wrap_content"
        android:layout_alignParentTop="true"
        android:layout_centerHorizontal="true"
        android:layout_marginTop="90dp"
        android:onClick="changeColor"
    />

</RelativeLayout>
```

The layout file defines a **RelativeLayout** containing an **AnalogClock**. The important part is the **onClick** attribute in the **AnalogClock** declaration.

```
android:onClick="changeColor"
```

This means that when the user presses the **AnalogClock** widget, the **changeColor** method in the activity class will be called. For a callback method like **changeColor** to work, it must have no return value and accept an argument of type **View**. The system will call this method and pass the widget that was pressed.

The **changeColor** method is part of the **MainActivity** class shown in Listing 5.3.

Listing 5.3: The MainActivity class in MulticolorClock

```
package com.example.multicolorclock;
import android.app.Activity;
import android.graphics.Color;
import android.os.Bundle;
import android.view.Menu;
import android.view.View;
import android.widget.AnalogClock;

public class MainActivity extends Activity {

    int counter = 0;
    int[] colors = { Color.BLACK, Color.BLUE, Color.CYAN,
            Color.DKGRAY, Color.GRAY, Color.GREEN, Color.LTGRAY,
            Color.MAGENTA, Color.RED, Color.WHITE, Color.YELLOW };

    @Override
    protected void onCreate(Bundle savedInstanceState) {
        super.onCreate(savedInstanceState);
        setContentView(R.layout.activity_main);
    }

    @Override
    public boolean onCreateOptionsMenu(Menu menu) {
        // Inflate the menu; this adds items to the action bar if it
```

```
        // is present.
        getMenuInflater().inflate(R.menu.menu_main, menu);
        return true;
    }

    public void changeColor(View view) {
        if (counter == colors.length) {
            counter = 0;
        }
        view.setBackgroundColor(colors[counter++]);
    }
}
```

Pay special attention to the **changeColor** method in the **MainActivity** class. When the user presses (or touches) the analog clock, this method will be called and receive the clock object. To change the clock's color, call its **setBackgroundColor** method, passing a color object. In Android, colors are represented by the **android.graphics.Color** class. The class has pre-defined colors that make creating color objects easy. These pre-defined colors include **Color.BLACK**, **Color.Magenta**, **Color.GREEN**, and others. The **MainActivity** class defines an array of **int**s that contains some of the pre-defined colors in **android.graphics.Color**.

```
int[] colors = { Color.BLACK, Color.BLUE, Color.CYAN,
        Color.DKGRAY, Color.GRAY, Color.GREEN, Color.LTGRAY,
        Color.MAGENTA, Color.RED, Color.WHITE, Color.YELLOW };
```

There is also a counter that points to the current index position of **colors**. The **changeColor** method inquiries the value of **counter** and changes it to zero if the value is equal to the array length. It then passes the pointed color to the **setBackgroundColor** method of the **AnalogClock**.

```
        view.setBackgroundColor(colors[counter++]);
```

The application is shown in Figure 5.1.

Figure 5.1: The MulticolorClock application

Touch the clock to change its color!

Implementing A Listener

As a second example, the GestureDemo application shows you how to implement the
View.OnTouchListener interface to handle the touch event. To make it simple, the
application only has one activity that contains a grid of cells that can be swapped. The
application is shown in Figure 5.2.

Figure 5.2: The GestureDemo application

Each of the images is an instance of the **CellView** class given in Listing 5.4. It simply extends **ImageView** and adds **x** and **y** fields to store the position in the grid.

Listing 5.4: The CellView class

```
package com.example.gesturedemo;
import android.content.Context;
import android.widget.ImageView;

public class CellView extends ImageView {
    int x;
    int y;

    public CellView(Context context, int x, int y) {
        super(context);
        this.x = x;
        this.y = y;
    }
}
```

There is no layout class for the activity as the layout is built programmatically. This is shown in the **onCreate** method of the **MainActivity** class in Listing 5.5.

Listing 5.5: The MainActivity class

```
package com.example.gesturedemo;
import android.app.Activity;
import android.graphics.drawable.Drawable;
import android.os.Bundle;
import android.view.Gravity;
import android.view.Menu;
import android.view.MotionEvent;
import android.view.View;
import android.view.View.OnTouchListener;
import android.view.ViewGroup;
import android.widget.ImageView;
import android.widget.LinearLayout;

public class MainActivity extends Activity {

    int rowCount = 7;
    int cellCount = 7;
    ImageView imageView1;
    ImageView imageView2;
    CellView[][] cellViews;
    int downX;
    int downY;
    boolean swapping = false;

    @Override
    protected void onCreate(Bundle savedInstanceState) {
        super.onCreate(savedInstanceState);

        LinearLayout root = new LinearLayout(this);
        LinearLayout.LayoutParams matchParent =
                new LinearLayout.LayoutParams(
                LinearLayout.LayoutParams.MATCH_PARENT,
                LinearLayout.LayoutParams.MATCH_PARENT);
        root.setOrientation(LinearLayout.VERTICAL);
        root.setGravity(Gravity.CENTER_VERTICAL);

        addContentView(root, matchParent);


        // create row
        cellViews = new CellView[rowCount][cellCount];
        LinearLayout.LayoutParams rowLayoutParams =
                new LinearLayout.LayoutParams(
                        LinearLayout.LayoutParams.MATCH_PARENT,
                        LinearLayout.LayoutParams.WRAP_CONTENT);

        ViewGroup.LayoutParams cellLayoutParams =
                new ViewGroup.LayoutParams(
                        ViewGroup.LayoutParams.WRAP_CONTENT,
                        ViewGroup.LayoutParams.WRAP_CONTENT);
```

```
        int count = 0;
        for (int i = 0; i < rowCount; i++) {
            CellView[] cellRow = new CellView[cellCount];
            cellViews[i] = cellRow;

            LinearLayout row = new LinearLayout(this);
            row.setLayoutParams(rowLayoutParams);
            row.setOrientation(LinearLayout.HORIZONTAL);
            row.setGravity(Gravity.CENTER_HORIZONTAL);
            root.addView(row);
            // create cells
            for (int j = 0; j < cellCount; j++) {
                CellView cellView = new CellView(this, j, i);
                cellRow[j] = cellView;
                if (count == 0) {
                    cellView.setImageDrawable(
                            getResources().getDrawable(
                                    R.drawable.image1));
                } else if (count == 1) {
                    cellView.setImageDrawable(
                            getResources().getDrawable(
                                    R.drawable.image2));
                } else {
                    cellView.setImageDrawable(
                            getResources().getDrawable(
                                    R.drawable.image3));
                }
                count++;
                if (count == 3) {
                    count = 0;
                }
                cellView.setLayoutParams(cellLayoutParams);
                cellView.setOnTouchListener(touchListener);
                row.addView(cellView);
            }
        }
    }

    @Override
    public boolean onCreateOptionsMenu(Menu menu) {
        getMenuInflater().inflate(R.menu.menu_main, menu);
        return true;
    }

    private void swapImages(CellView v1, CellView v2) {
        Drawable drawable1 = v1.getDrawable();
        Drawable drawable2 = v2.getDrawable();
        v1.setImageDrawable(drawable2);
        v2.setImageDrawable(drawable1);
    }

    OnTouchListener touchListener = new OnTouchListener() {
        @Override
```

```java
public boolean onTouch(View v, MotionEvent event) {
    CellView cellView = (CellView) v;

    int action = event.getAction();
    switch (action) {
    case (MotionEvent.ACTION_DOWN):
        downX = cellView.x;
        downY = cellView.y;
        return true;
    case (MotionEvent.ACTION_MOVE):
        if (swapping) {
            return true;
        }
        float x = event.getX();
        float y = event.getY();
        int w = cellView.getWidth();
        int h = cellView.getHeight();
        if (downX < cellCount - 1
                && x > w && y >= 0 && y <= h) {
            // swap with right cell
            swapping = true;
            swapImages(cellView,
                    cellViews[downY][downX + 1]);
        } else if (downX > 0 && x < 0
                && y >=0 && y <= h) {
            // swap with left cell
            swapping = true;
            swapImages(cellView,
                    cellViews[downY][downX - 1]);
        } else if (downY < rowCount - 1
                && y > h && x >= 0 && x <= w) {
            // swap with cell below
            swapping = true;
            swapImages(cellView,
                    cellViews[downY + 1][downX]);
        } else if (downY > 0 && y < 0
                && x >= 0 && x <= w) {
            // swap with cell above
            swapping = true;
            swapImages(cellView,
                    cellViews[downY - 1][downX]);
        }
        return true;
    case (MotionEvent.ACTION_UP):
        swapping = false;
        return true;
    default:
        return true;
    }
}
};
}
```

The **MainActivity** class contains a **View.OnTouchListener** called **touchListener** that will be attached to every single **CellView** in the grid. The **OnTouchListener** interface has an **onTouch** method that must be implemented. Here is the signature of **onTouch**.

```
public boolean onTouch(View view, MotionEvent event)
```

The method should return **true** if it has consumed the event, which means that the event should not propagate to other views. Otherwise, it should return **false**.

A single touch action by the user causes the **onTouch** method to be called several times. When the user touches the view, the method is called. When the user moves his/her finger, **onTouch** is called. Likewise, **onTouch** is called when the user lifts his/her finger. The second argument to **onTouch**, a **MotionEvent**, contains the information about the event. You can inquire what type of action is triggering the event by calling the **getAction** method on the **MotionEvent**.

```
int action = event.getAction();
```

The return value will be one of the static final **int**s defined in the **MotionEvent** class. For this application we are interested in **MotionEvent.ACTION_DOWN**, **MotionEvent.ACTION_MOVE**, and **MotionEvent.ACTION_UP**. When the user touches the view, the **getAction** method returns a **MotionEvent.ACTION_DOWN**. The code simply stores the location of the event to the **x** and **y** fields and returns **true**.

```
case (MotionEvent.ACTION_DOWN):
    downX = cellView.x;
    downY = cellView.y;
    return true;
```

If the user moves his/her finger to a neighboring cell, the touch action will return a **MotionEvent.ACTION_MOVE** and you need to swap the images of the original cell and the destination cell and set the **swapping** field to true. This would prevent another swapping before the finger is lifted.

Finally, when the user lifts his/her finger, the swapping field is set to false to enable another swapping.

The layout for the activity is built dynamically in the **onCreate** method of the activity class. Each **CellView** is passed the **OnTouchListener** so that the listener will handle the **CellView**'s touch event.

```
cellView.setOnTouchListener(touchListener);
```

Summary

In this chapter you learned the basics of Android event handling and how to write listeners by implementing a nested interface in the **View** class. You have also learned to use the shortcut for handling the click event.

Chapter 6
The Action Bar

The action bar is a rectangular window area that contains the application icon, application name, menus and other navigation buttons. The action bar normally appears at the top of the window.

This chapter explains how to decorate the action bar on Android with API level 11 (Android 3.0) or higher.

Overview

The action bar is represented by the **android.app.ActionBar** class. It should look familiar to any Android user. Figure 6.1 shows the action bar of the Messaging application and Figure 6.2 shows that of Calendar.

Figure 6.1: The action bar of Messaging

Figure 6.2: The action bar of Calendar

The application icon and name on the left of the action bar are there by default. They are both optional and no programming is needed to display them. The system will use the values of the application element's **android:icon** and **android:label** attributes in the manifest. Other item types, such as navigation tabs or an options menu, have to be added using code.

The rightmost icon on the action bar (the one with three little dots) is called the (action) overflow button. When pressed, the overflow button displays action items that may do an action if selected. Important action items can be configured to display directly on the action bar instead of hidden in the overflow. An action item shown on the action bar is called an action button. An action button can have an icon, a label, or both. For example, the action bar in Figure 6.1 contains two action buttons, New Message and Search. The New Message action button has both an icon and a label. The Search action button only has an icon. The action bar in Figure 6.2 also contains two action buttons.

In Android 3.0 or higher, the action bar is shown automatically. You can hide the action bar if you wish by adding this code in the **onCreate** method of your activity.

```
getActionBar().hide();
```

To show a hidden action bar, call the **show** method:

```
getActionBar().hide();
```

The next sections show how to add action items and drop-down navigation.

Note
You can download Android's icon pack that contains icons for your action bar from this site.

```
http://developer.android.com/downloads/design/
Android_Design_Icons_20131106.zip
```

Adding Action Items

To add action items to the action overflow, follow these two steps.

1. Create a menu in an xml file and save it under the **res/menu** directory. Android Studio will add a field to your **R.menu** class so that you can load the menu in your application. The field name is the same as the XML file minus the extension. If the XML file is called **main_activity_menu.xml**, for example, the field will be called **main_activity_menu**.
2. In your activity class, override the **onCreateOptionsMenu** method and call **getMenuInflater().inflate()**, passing the menu to be loaded and the menu passed to the method, like this.

```
@Override
public boolean onCreateOptionsMenu(Menu menu) {
    getMenuInflater().inflate(R.menu.main, menu);
    return true;
}
```

An action item that does nothing is useless. For an action item to be able to respond to a action item being selected, you must override the **onOptionsItemSelected** method in your activity class. This method is called every time an item menu is selected and the system will pass the **MenuItem** selected. The signature of the method is as follows.

```
public boolean onOptionsItemSelected(MenuItem item);
```

You can find out which menu item was selected by calling the **getItemId** on the **MenuItem** argument. Normally you would use a **switch** statement like this:

```
switch (item.getItemId()) {
    case R.id.action_1:
        // do something
        return true;
    case R.id.action_2:
        // do something else
        return true;
...
```

Now that you know the theory, let's add some item actions. The ActionBarDemo application shows how to do it. It adds three action items to the action bar.

As usual, let's start with the manifest, which for this example is shown in Listing 6.1.

Listing 6.1: The manifest for ActionBarDemo

```xml
<?xml version="1.0" encoding="utf-8"?>
<manifest xmlns:android="http://schemas.android.com/apk/res/android"
    package="com.example.actionbardemo"
    android:versionCode="1"
    android:versionName="1.0" >

    <uses-sdk
        android:minSdkVersion="11"
        android:targetSdkVersion="18" />

    <application
        android:allowBackup="true"
        android:icon="@drawable/ic_launcher"
        android:label="@string/app_name"
        android:theme="@style/AppTheme" >
        <activity
            android:name="com.example.actionbardemo.MainActivity"
            android:label="@string/app_name" >
            <intent-filter>
                <action android:name="android.intent.action.MAIN"/>
                <category
android:name="android.intent.category.LAUNCHER" />
            </intent-filter>
        </activity>
    </application>
</manifest>
```

It is good practice to list action names in a resource file. Listing 6.2 shows the **strings.xml** file that contains three strings for the action items, **action_capture**, **action_profile** and **action_about**.

Listing 6.2: The res/values/strings.xml

```xml
<?xml version="1.0" encoding="utf-8"?>
<resources>
    <string name="app_name">ActionBarDemo</string>
    <string name="action_capture">Capture</string>
    <string name="action_profile">Profile</string>
    <string name="action_about">About</string>
    <string name="hello_world">Hello world!</string>
</resources>
```

Next, create an XML file under **res/menu**. If you used Android Studio to create the Android application, one has been created for you. You just need to add **item** elements to it. Listing 6.3 shows the menu for the action items.

Listing 6.3: The res/menu/menu_main.xml

```xml
<menu xmlns:android="http://schemas.android.com/apk/res/android">
    <item
        android:id="@+id/action_capture"
        android:orderInCategory="100"
        android:showAsAction="ifRoom|withText"
        android:icon="@drawable/icon1"
        android:title="@string/action_capture"/>

    <item
        android:id="@+id/action_profile"
        android:orderInCategory="200"
        android:showAsAction="ifRoom|withText"
        android:icon="@drawable/icon2"
        android:title="@string/action_profile"/>

    <item
        android:id="@+id/action_about"
        android:orderInCategory="50"
        android:showAsAction="never"
        android:title="@string/action_about"/>
</menu>
```

The **item** element may have any of these attributes.

- **android:id**. A unique identifier to refer to the action item in the program.
- **android:orderInCategory**. The order number for this item. An item with a smaller number will be shown before items with larger numbers.
- **android:icon**. The icon for this action item if it is shown as an action button (directly on the action bar).
- **android:title**. The action label.
- **android:showAsAction**. The value can be one or a combination of these values: **ifRoom**, **never**, **withText**, **always**, and **collapseActionView**. Populating this attribute with **never** indicates that this item will never be shown on the action bar directly. On the other hand, **always** forces the system to always display this item as an action button. However, be cautious when using this value as if there is not enough room on the action bar, what will be displayed will be unpredictable. Instead, use **ifRoom** to display an item as an action button if there is room. The **withText** value will display this item with a label if this item is being displayed as an action button.

The complete list of attributes for the **item** element can be found here.

```
http://developer.android.com/guide/topics/resources/menu-
resource.html
```

Finally, Listing 6.4 presents the **MainActivity** class for the application.

Listing 6.4: The MainActivity class

```java
package com.example.actionbardemo;
import android.app.Activity;
import android.app.AlertDialog;
import android.os.Bundle;
```

```
import android.view.Menu;
import android.view.MenuItem;

public class MainActivity extends Activity {

    @Override
    protected void onCreate(Bundle savedInstanceState) {
        super.onCreate(savedInstanceState);
        setContentView(R.layout.activity_main);
    }

    @Override
    public boolean onCreateOptionsMenu(Menu menu) {
        getMenuInflater().inflate(R.menu.menu_main, menu);
        return true;
    }

    @Override
    public boolean onOptionsItemSelected(MenuItem item) {
        // Handle presses on the action bar items
        switch (item.getItemId()) {
            case R.id.action_profile:
                showAlertDialog("Profile", "You selected Profile");
                return true;
            case R.id.action_capture:
                showAlertDialog("Settings",
                        "You selected Settings");
                return true;
            case R.id.action_about:
                showAlertDialog("About", "You selected About");
                return true;
            default:
                return super.onOptionsItemSelected(item);
        }
    }

    private void showAlertDialog(String title, String message) {
        AlertDialog alertDialog = new
                AlertDialog.Builder(this).create();
        alertDialog.setTitle(title);
        alertDialog.setMessage(message);
        alertDialog.show();
    }
}
```

Noticed that the activity class overrides the **onOptionsItemSelected** method? Selecting an item will invoke the **showAlertDialog** method that shows an **AlertDialog**.

Figure 6.3 shows three action items in ActionBarDemo. Two of the items are displayed as action buttons.

Figure 6.3: The ActionBarDemo application

Adding Dropdown Navigation

A dropdown list can be used as a navigation mode. The visual difference between a dropdown list and an options menu is that a dropdown list always displays the selected item on the action bar and hide the other options. On the other hand, an options menu may hide all of the items or show all or some of them as action buttons. Figure 6.4 shows dropdown navigation in Calendar.

Figure 6.4: Drop down navigation in Calendar

To add drop-down navigation to the action bar, follow these three steps.

1. Declare a string array in your **strings.xml** file under **res/values**.
2. In your activity class, add an implementation of **ActionBar.OnNavigationListener** to respond to item selection.
3. Create a **SpinnerAdapter** in the **onCreate** method of your activity, pass **ActionBar.NAVIGATION_MODE_LIST** to the **ActionBar**'s **setNavigationMode** method, and pass the **SpinnerAdapter** and **OnNavigationListener** to the **ActionBar**'s **setListNavigationCallbacks** method.

```
SpinnerAdapter spinnerAdapter =
        ArrayAdapter.createFromResource(this,
        R.array.colors,
        android.R.layout.simple_spinner_dropdown_item);
ActionBar actionBar = getActionBar();
actionBar.setNavigationMode(
        ActionBar.NAVIGATION_MODE_LIST);
actionBar.setListNavigationCallbacks(spinnerAdapter,
        onNavigationListener);
```

As an example, the DropDownNavigationDemo application shows how to add dropdown navigation to the action bar. The application adds a list of five colors to the action bar. Selecting a color changes the window background color with the selected color.

Listing 6.5 shows the manifest for the application.

Listing 6.5: The DropDownNavigationDemo manifest

```xml
<?xml version="1.0" encoding="utf-8"?>
<manifest xmlns:android="http://schemas.android.com/apk/res/android"
    package="com.example.dropdownnavigationdemo"
    android:versionCode="1"
    android:versionName="1.0" >

    <uses-sdk
        android:minSdkVersion="14"
        android:targetSdkVersion="18" />

    <application
        android:allowBackup="true"
        android:icon="@drawable/ic_launcher"
        android:label="@string/app_name"
        android:theme="@style/AppTheme" >
        <activity
android:name="com.example.dropdownnavigationdemo.MainActivity"
            android:label="@string/app_name"
            android:theme="@style/MyTheme">
            <intent-filter>
                <action android:name="android.intent.action.MAIN"/>
                <category
android:name="android.intent.category.LAUNCHER" />
            </intent-filter>
        </activity>
    </application>
</manifest>
```

Listing 6.6 shows a **string-array** element that will be used to populate the drop-down. There are five items in the array.

Listing 6.6: The res/values/strings.xml file

```xml
<?xml version="1.0" encoding="utf-8"?>
<resources>
    <string name="app_name">DropDownNavigationDemo</string>
    <string name="action_settings">Settings</string>
    <string name="hello_world">Hello world!</string>

    <string-array name="colors">
        <item>White</item>
        <item>Red</item>
        <item>Green</item>
        <item>Blue</item>
        <item>Yellow</item>
    </string-array>
</resources>
```

Listing 6.7 shows the **MainActivity** class for the application.

Listing 6.7: The MainActivity class

```
package com.example.dropdownnavigationdemo;
import android.app.ActionBar;
import android.app.ActionBar.OnNavigationListener;
import android.app.Activity;
import android.graphics.Color;
import android.os.Bundle;
import android.view.Menu;
import android.widget.ArrayAdapter;
import android.widget.SpinnerAdapter;

public class MainActivity extends Activity {

    @Override
    protected void onCreate(Bundle savedInstanceState) {
        super.onCreate(savedInstanceState);
        setContentView(R.layout.activity_main);
        SpinnerAdapter spinnerAdapter =
                ArrayAdapter.createFromResource(this,
                R.array.colors,
                android.R.layout.simple_spinner_dropdown_item);
        ActionBar actionBar = getActionBar();
        actionBar.setNavigationMode(
                ActionBar.NAVIGATION_MODE_LIST);
        actionBar.setListNavigationCallbacks(spinnerAdapter,
                onNavigationListener);
    }

    @Override
    public boolean onCreateOptionsMenu(Menu menu) {
        getMenuInflater().inflate(R.menu.menu_main, menu);
        return true;
    }

    OnNavigationListener onNavigationListener = new
            OnNavigationListener() {
        @Override
        public boolean onNavigationItemSelected(
                int position, long itemId) {
            String[] colors = getResources().
                    getStringArray(R.array.colors);
            String selectedColor = colors[position];

            getWindow().getDecorView().setBackgroundColor(
                    Color.parseColor(selectedColor));
            return true;
        }
    };
}
```

Figure 6.5 shows the dropdown navigation.

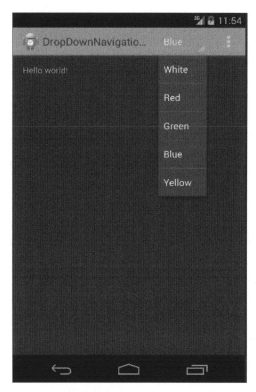

Figure 6.5: Dropdown navigation on the action bar

Note that the action bar has been styled using the **styles.xml** file in Listing 6.8.

Listing 6.8: The res/values/styles.xml file

```
<resources>
    <style name="AppBaseTheme" parent="android:Theme.Light">
    </style>

    <style name="AppTheme" parent="AppBaseTheme">
    </style>

    <style name="MyTheme"
            parent="@android:style/Widget.Holo.Light">
        <item name="android:actionBarStyle">@style/MyActionBar</item>
    </style>

    <style name="MyActionBar"
parent="@android:style/Widget.Holo.Light.ActionBar.Solid.Inverse">
        <item
name="android:background">@android:color/holo_blue_bright</item>
    </style>
</resources>
```

For more information on styling UI components, see Chapter 10, "Styles and Themes."

Going Back Up

You can set the application icon and activity label in the action bar of an activity so that the application will go one level back up when the icon is pressed. Figure 6.6 shows an action bar whose **displayHomeAsUp** property is set to **true** (indicated by the left arrow to the left of the icon). Compare this with the action bar in Figure 6.7 where its **displayHomeAsUp** property is set to **false**.

Figure 6.6: displayHomeAsUp set to true

Figure 6.7: displayHomeAsUp set to false

To enable **displayHomeAsUp**, you need to set the **parentActivityName** element in the activity declaration in the Android manifest:

```
<activity android:name="com.example.d1.ShowContactActivity"
    android:parentActivityName=".MainActivity">
</activity>
```

You must also leave the **displayHomeAsUpEnabled** property of the action bar to its default value (**true**). Setting it to **false**, as shown in the code below, will disable it.

```
getActionBar().setDisplayHomeAsUpEnabled(false);
```

Summary

The action bar provides a space for the application icon, application name and navigation modes. This chapter showed how to add action items and dropdown navigation to the action bar.

Chapter 7
Menus

Menus are a common feature in many graphical user interface (GUI) systems. Their primary role is to provide shortcuts to certain actions.

This chapter looks at Android menus closely and provides three sample applications.

Overview

Pre-3.0 Android devices shipped with a (hardware) button for showing menus in the active application. Starting from Android 3.0, the action bar is the recommended way of achieving the same thing, in effect making a hardware Menu button redundant. With the hardware menu button gone, "soft" menus have become even more important than ever.

There are three types of menus in Android:

- Options menu
- Context menu
- Popup menu

The options menu is the type of menu you normally incorporate in the action bar, as you have seen in Chapter 6, "The Action Bar." In this chapter you will look at the options menu more closely and learn about the other two. Thankfully, no matter what kind of menu you're using in your app, you use the same API. And, yes, you can use different types of menus in the same application.

Like many other things in Android, menus can be defined declaratively or programmatically. The first method offers more flexibility than the second because it allows you to change menu items using a text editor. Doing so programmatically, on the other hand, would require you to change your program and recompile every time you need to edit your menu.

Here are the three things you need to do when working with options and context menus.

1. Create a menu in an xml file and save it under the **res/menu** directory.
2. In your activity class, override either **onCreateOptionsMenu** or **onCreateContextMenu**, depending on the menu type. Then, in the overridden method, call **getMenuInflater().inflate()**, passing the menu to be used.
3. In your activity class, override either **onOptionsItemSelected** or **onContextItemSelected**, depending on the menu type.

Popup menus are a bit different. To work with them, do the following:

1. Create a menu in an xml file and save it under the **res/menu** directory.
2. In your activity class, create a **PopupMenu** object and a **PopupMenu.OnMenuItemClickListener** object. In the listener class you define a method that handles the click event that occurs when one of the popup menu items is selected.

The Menu File

To create a menu declaratively, start by creating an XML file and place it under the **res/menu** directory. The XML file must have the following structure.

```
<menu xmlns:android="http://schemas.android.com/apk/res/android">
    <group>...</group>
    <group>...</group>
    ...
    <item>...</item>
    <item>...</item>
    ...
</menu>
```

The root element is **menu** and it can contain any number of **group** and **item** elements. The **group** element represents a menu group and the **item** element represents a menu item.

For every menu file you create, Android Studio will add a field to your **R.menu** class so that you can load the menu in your application. The field name is the same as the XML file minus the extension. If the XML file is called **main_activity_menu.xml**, for example, the field in **R.menu** will be called **main_activity_menu**.

Note
The **R** class was explained in Chapter 1, "Getting Started."

The Options Menu

The OptionsMenuDemo application is a simple application that uses an options menu in its action bar. It is similar to the application that demonstrates the action bar in Chapter 6, "The Action Bar."

The manifest (**AndroidManifest.xml** file) for this application is shown in Listing 7.1.

Listing 7.1: The manifest for OptionsMenuDemo

```
<?xml version="1.0" encoding="utf-8"?>
<manifest xmlns:android="http://schemas.android.com/apk/res/android"
    package="com.example.optionsmenudemo"
    android:versionCode="1"
    android:versionName="1.0" >
```

```
<uses-sdk
    android:minSdkVersion="18"
    android:targetSdkVersion="18" />

<application
    android:allowBackup="true"
    android:icon="@drawable/ic_launcher"
    android:label="@string/app_name"
    android:theme="@style/AppTheme" >
    <activity
        android:name="com.example.optionsmenudemo.MainActivity"
        android:label="@string/app_name" >
        <intent-filter>
            <action android:name="android.intent.action.MAIN"/>
            <category
android:name="android.intent.category.LAUNCHER"/>
        </intent-filter>
    </activity>
</application>
</manifest>
```

The manifest declares an activity, whose class is called **MainActivity**.

The menu for this application is defined in the **res/menu/options_menu.xml** file in Listing 7.2. It has three menu items.

Listing 7.2: The options_menu.xml File

```
<menu xmlns:android="http://schemas.android.com/apk/res/android">
    <item
        android:id="@+id/action_capture"
        android:orderInCategory="100"
        android:showAsAction="ifRoom|withText"
        android:icon="@drawable/icon1"
        android:title="@string/action_capture"/>

    <item
        android:id="@+id/action_profile"
        android:orderInCategory="200"
        android:showAsAction="ifRoom|withText"
        android:icon="@drawable/icon2"
        android:title="@string/action_profile"/>

    <item
        android:id="@+id/action_about"
        android:orderInCategory="50"
        android:showAsAction="never"
        android:title="@string/action_about"/>
</menu>
```

As explained in Chapter 4, "Layouts," the plus sign in an **id** attribute indicates that the identifier is being added with the declaration.

The titles for the menu items reference the strings defined in the **res/values/strings.xml** file in Listing 7.3.

Listing 7.3: strings.xml for OptionsMenuDemo

```xml
<?xml version="1.0" encoding="utf-8"?>
<resources>
    <string name="app_name">OptionsMenuDemo</string>
    <string name="action_capture">Capture</string>
    <string name="action_profile">Profile</string>
    <string name="action_about">About</string>
    <string name="hello_world">Hello world!</string>
</resources>
```

The activity class for the application, the **MainActivity** class, is shown in Listing 7.4.

Listing 7.4: MainActivity for OptionsMenuDemo

```java
package com.example.optionsmenudemo;
import android.app.Activity;
import android.app.AlertDialog;
import android.os.Bundle;
import android.view.Menu;
import android.view.MenuItem;

public class MainActivity extends Activity {
    @Override
    protected void onCreate(Bundle savedInstanceState) {
        super.onCreate(savedInstanceState);
        setContentView(R.layout.activity_main);
    }

    @Override
    public boolean onCreateOptionsMenu(Menu menu) {
        getMenuInflater().inflate(R.menu.options_menu, menu);
        return true;
    }

    @Override
    public boolean onOptionsItemSelected(MenuItem item) {
        // Handle click on menu items
        switch (item.getItemId()) {
            case R.id.action_profile:
                showAlertDialog("Profile", "You selected Profile");
                return true;
            case R.id.action_capture:
                showAlertDialog("Settings",
                        "You selected Settings");
                return true;
            case R.id.action_about:
                showAlertDialog("About", "You selected About");
                return true;
            default:
                return super.onOptionsItemSelected(item);
        }
    }

    private void showAlertDialog(String title, String message) {
```

```
        AlertDialog alertDialog = new
                AlertDialog.Builder(this).create();
        alertDialog.setTitle(title);
        alertDialog.setMessage(message);
        alertDialog.show();
    }
}
```

To use the options menu you need to override the **onCreateOptionsMenu** and **onOptionsItemSelected** methods. The **onCreateOptionsMenu** method is called when the activity is built. You should call the menu inflater and inflate your menu here. In addition, the **onOptionsItemSelected** method handles menu item selection.

Note that the options menu is integrated with the activity so that you do not need to create your own listener to handle item selection.

If you run the application, you will see an activity like the one in Figure 7.1. Take a look at the action bar and try selecting one of the menu items. Every time you select a menu item, an **AlertDialog** will be shown to notify what you have selected.

Figure 7.1: OptionsMenuDemo

In Figure 7.1 the buttons on the action bar are rendered without text because the application is running in a device with a low-resolution screen. If you run it in a device with a higher resolution screen, you may see text to the right of each button.

The Context Menu

The ContextMenuDemo application shows how you can use a context menu in your application. The main activity of the application features an image button that you can long-press to display a context menu.

The **AndroidManifest.xml** file for this application is printed in Listing 7.5.

Listing 7.5: AndroidMenifest.xml for ContextMenuDemo

```xml
<?xml version="1.0" encoding="utf-8"?>
<manifest xmlns:android="http://schemas.android.com/apk/res/android"
    package="com.example.contextmenudemo"
    android:versionCode="1"
    android:versionName="1.0" >

    <uses-sdk
        android:minSdkVersion="18"
        android:targetSdkVersion="18" />

    <application
        android:allowBackup="true"
        android:icon="@drawable/ic_launcher"
        android:label="@string/app_name"
        android:theme="@style/AppTheme" >
        <activity
            android:name="com.example.contextmenudemo.MainActivity"
            android:label="@string/app_name" >
            <intent-filter>
                <action android:name="android.intent.action.MAIN"/>
                <category
android:name="android.intent.category.LAUNCHER"/>
            </intent-filter>
        </activity>
    </application>
</manifest>
```

The **context_menu.xml** file in Listing 7.6 is a menu file that defines menu items for the context menu used in the application.

Listing 7.6: context_menu.xml for ContextMenuDemo

```xml
<menu xmlns:android="http://schemas.android.com/apk/res/android">
    <item
        android:id="@+id/action_rotate"
        android:title="@string/action_rotate"/>
    <item
        android:id="@+id/action_resize"
        android:title="@string/action_resize"/>
</menu>
```

The menu file defines two menu items, whose titles get their values from the **res/values/strings.xml** file in Listing 7.7.

Listing 7.7: strings.xml for ContextMenuDemo

```xml
<?xml version="1.0" encoding="utf-8"?>
<resources>
    <string name="app_name">ContextMenuDemo</string>
    <string name="action_settings">Settings</string>
    <string name="action_rotate">Rotate</string>
    <string name="action_resize">Resize</string>
    <string name="hello_world">Hello world!</string>
</resources>
```

Finally, Listing 7.8 shows the **MainActivity** class for the application. There are two methods that you need to override to use a context menu, **onCreateContextMenu** and **onContextItemSelected**. The **onCreateContextMenu** method is called when the activity is built. You should inflate your menu here.

The **onContextItemSelected** method is called every time a menu item in the context menu is selected.

Listing 7.8: MainActivity for ContextMenuDemo

```java
package com.example.contextmenudemo;
import android.app.Activity;
import android.app.AlertDialog;
import android.os.Bundle;
import android.view.ContextMenu;
import android.view.ContextMenu.ContextMenuInfo;
import android.view.MenuInflater;
import android.view.MenuItem;
import android.view.View;
import android.widget.ImageButton;

public class MainActivity extends Activity {
    @Override
    protected void onCreate(Bundle savedInstanceState) {
        super.onCreate(savedInstanceState);
        setContentView(R.layout.activity_main);
        ImageButton imageButton = (ImageButton)
            findViewById(R.id.button1);
        registerForContextMenu(imageButton);
    }

    @Override
    public void onCreateContextMenu(ContextMenu menu, View v,
            ContextMenuInfo menuInfo) {
        super.onCreateContextMenu(menu, v, menuInfo);
        MenuInflater inflater = getMenuInflater();
        inflater.inflate(R.menu.context_menu, menu);
    }
    @Override
    public boolean onContextItemSelected(MenuItem item) {
        switch (item.getItemId()) {
            case R.id.action_rotate:
                showAlertDialog("Rotate", "You selected Rotate ");
                return true;
```

```
        case R.id.action_resize:
            showAlertDialog("Resize", "You selected Resize");
            return true;
        default:
            return super.onContextItemSelected(item);
    }
}

private void showAlertDialog(String title, String message) {
    AlertDialog alertDialog = new
            AlertDialog.Builder(this).create();
    alertDialog.setTitle(title);
    alertDialog.setMessage(message);
    alertDialog.show();
}
}
```

Figure 7.2 shows the application. If you press (or click) the image button long enough, it will show the context menu. Note that the image comes from the Android system as is defined in the layout file.

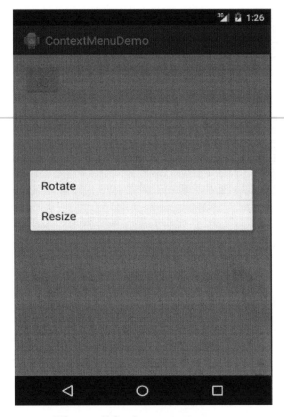

Figure 7.2: A context menu

The Popup Menu

A popup menu is associated with a view and is shown every time an event occurs to the view. The **PopupMenuDemo** application shows how to use a popup menu. It uses a button that displays a popup menu when it is clicked. Listing 7.9 shows the **AndroidManifest.xml** file for **PopupMenuDemo**.

Listing 7.9: AndroidMenifest.xml for PopupMenuDemo

```
<?xml version="1.0" encoding="utf-8"?>
<manifest xmlns:android="http://schemas.android.com/apk/res/android"
    package="com.example.popupmenudemo"
    android:versionCode="1"
    android:versionName="1.0" >

    <uses-sdk
        android:minSdkVersion="18"
        android:targetSdkVersion="18" />

    <application
        android:allowBackup="true"
        android:icon="@drawable/ic_launcher"
        android:label="@string/app_name"
        android:theme="@style/AppTheme" >
        <activity
            android:name="com.example.popupmenudemo.MainActivity"
            android:label="@string/app_name" >
            <intent-filter>
                <action android:name="android.intent.action.MAIN"/>
                <category
android:name="android.intent.category.LAUNCHER"/>
            </intent-filter>
        </activity>
    </application>
</manifest>
```

The manifest in Listing 7.9 is a standard XML file that you've seen many times. It has one activity with a button that will activate the menu shown in Listing 7.10.

Listing 7.10: popup_menu.xml for PopupMenuDemo

```
<menu xmlns:android="http://schemas.android.com/apk/res/android">
    <item
        android:id="@+id/action_delete"
        android:title="@string/action_delete"/>
    <item
        android:id="@+id/action_copy"
        android:title="@string/action_copy"/>
</menu>
```

The menu in Listing 7.10 has two menu items. The titles for the items refer to the strings defined in the **res/values/strings.xml** file in Listing 7.11.

Listing 7.11: strings.xml for PopupMenuDemo

```xml
<?xml version="1.0" encoding="utf-8"?>
<resources>
    <string name="app_name">PopupMenuDemo</string>
    <string name="action_settings">Settings</string>
    <string name="action_delete">Delete</string>
    <string name="action_copy">Copy</string>
    <string name="show_menu">Show Popup</string>
</resources>
```

Finally, Listing 7.12 shows the **MainActivity** class for the application.

Listing 7.12: MainActivity for PopupMenuDemo

```java
package com.example.popupmenudemo;
import android.app.Activity;
import android.os.Bundle;
import android.util.Log;
import android.view.MenuItem;
import android.view.View;
import android.widget.Button;
import android.widget.PopupMenu;

public class MainActivity extends Activity {

    PopupMenu popupMenu;
    PopupMenu.OnMenuItemClickListener menuItemClickListener;

    @Override
    protected void onCreate(Bundle savedInstanceState) {
        super.onCreate(savedInstanceState);
        setContentView(R.layout.activity_main);
        menuItemClickListener =
                new PopupMenu.OnMenuItemClickListener() {
            @Override
            public boolean onMenuItemClick(MenuItem item) {
                switch (item.getItemId()) {
                case R.id.action_delete:
                    Log.d("menu", "Delete clicked");
                    return true;
                case R.id.action_copy:
                    Log.d("menu", "Copy clicked");
                    return true;
                default:
                    return false;
                }
            }
        };
        Button button = (Button) findViewById(R.id.button1);
        popupMenu = new PopupMenu(this, button);
        popupMenu.setOnMenuItemClickListener(menuItemClickListener);
        popupMenu.inflate(R.menu.popup_menu);
    }
```

```
public void showPopupMenu(View view) {
    popupMenu.show();
}
}
```

Unlike the options menu and context menu, the popup menu requires that you create a menu object and a listener object for handling item selection.

In the **onCreate** method of **MainActivity**, you create a **PopupMenu** object and a **PopupMenu.OnMenuItemClickListener** object. You then pass the listener to the **PopupMenu**. The listener class handles menu item clicks.

The **showPopupMenu** method in **MainActivity** is associated with the button using the **onClick** attribute of the button in the main activity layout file. The method shows the popup menu.

Figure 7.3 shows the popup menu that displays when the button is clicked.

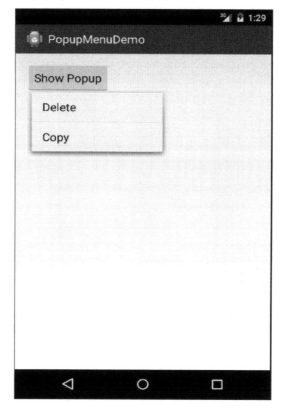

Figure 7.3: A popup menu

Summary

In this chapter you learned how to use menus to provide shortcuts to certain actions. There are three types of menus in Android, options menus, context menus, and popup menus.

Chapter 8
ListView

A **ListView** is a view for showing a scrollable list of items, which may come from a list adapter or an array adapter. Selecting an item in a **ListView** triggers an event for which you can write a listener.

If an activity contains only one view that is a **ListView**, you can extend **ListActivity** instead of **Activity** as your activity class. Using **ListActivity** is convenient as it comes with a number of useful features.

This chapter shows how you can use the **ListView** and **ListActivity** as well as create a custom **ListAdapter** and style a **ListView** in three sample applications.

Overview

Technically, **android.widget.ListView**, the template for creating a **ListView**, is a descendant of the **View** class. You can use it the same way you would other views. What makes **ListView** a bit tricky to use is the fact that you have to obtain a data source for it in the form of a **ListAdapter**. The **ListAdapter** also supply the layout for each item on the **ListView**, so the **ListAdapter** really plays a very important role in the life of a **ListView**.

The **android.widget.ListAdapter** interface is a subinterface of **android.widget.Adapter**. The close relatives of this interface are shown in Figure 8.1.

Creating a **ListAdapter** is explained in the next section, "Creating a ListAdapter." Once you have a **ListAdapter**, you can pass it to a **ListView**'s **setAdapter** method:

```
listView.setAdapter(listAdapter);
```

You can also write a listener that implements **AdapterView.OnItemClickListener** and pass it to the **ListView**'s **setOnItemClickListener** method. The listener will be notified every time a list item is selected and you can write code to handle it, like so.

```
listView.setOnItemClickListener(new
        AdapterView.OnItemClickListener() {
    @Override
    public void onItemClick(AdapterView<?> parent, final View view,
            int position, long id) {
        // handle item
    });
```

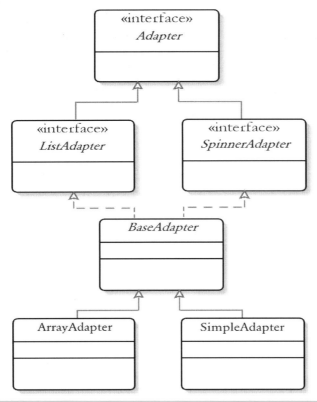

Figure 8.1: The parents and implementations of ListAdapter

Creating A ListAdapter

As mentioned in the previous section, the trickiest part of using a **ListView** is creating a data source for it. You need a **ListAdapter** and as you can see in Figure 8.1 you have at least two implementations of **ListAdapter** that you can use.

One of the concrete implementations of **ListAdapter** is the **ArrayAdapter** class. An **ArrayAdapter** is backed by an array of objects. The string returned by the **toString** method of each object is used to populate each item in the **ListView**.

The **ArrayAdapter** class offers several constructors. All of them require that you pass a **Context** and a resource identifier that points to a layout that contains a **TextView**. This is because each item in a **ListView** is a **TextView**. These are some of the constructors in the **ArrayAdapter** class.

```
public ArrayAdapter(android.content.Context context, int resourceId)

public ArrayAdapter(android.content.Context context, int resourceId,
        T[] objects)
```

If you do not pass an object array to a constructor, you will have to pass one later. As for the resource identifier, Android provides some pre-defined layouts for a **ListAdapter**. The identifiers to these layouts can be found in the **android.R.layout** class. For example, you can create an **ArrayAdapter** using this code snippet in your activity.

```
ArrayAdapter<String> adapter = new ArrayAdapter<String>(this,
        android.R.layout.simple_list_item_1, objects);
```

Using **android.R.layout.simple_list_item_1** will create a **ListView** with the simplest layout where the text for each item is printed in black. Alternatively, you can use **android.R.layout.simple_expandable_list_item_1**. However, you probably want to create your own layout and pass it to the constructor, instead. This way you would have more control over the look and feel of your **ListView**.

Most of the time you can use a string array as the data source for your **ListView**. You can create a string array programmatically or declaratively. Doing it programmatically is simple and you do not have to deal with an external resource:

```
String[] objects = { "item1", "item2","item-n" };
```

The disadvantage of this approach is that updating the array would require you to recompile your class. Creating a string array declaratively, on the other hand, gives you more flexibility as you can easily edit the elements.

To create a string array declaratively, start by creating a **string-array** element in your **strings.xml** file under **res/values**. For example, the following is a **string-array** named **players**.

```
<string-array name="players">
    <item>Player 1</item>
    <item>Player 2</item>
    <item>Player 3</item>
    <item>Player 4</item>
</string-array>
```

When you save the **strings.xml** file, Android Studio will update your **R** generated class and add a static final class named **array**, if none exists, as well as add a resource identifier for the **string-array** element to the **array** class. As a result, you now have this resource identifier to access your string array from your code:

```
R.array.players
```

To convert the user-defined string array to a Java string array, use this code.

```
String[] values = getResources().getStringArray(R.array.players);
```

You can then use this string array to create an **ArrayAdapter**.

Using A ListView

The ListViewDemo1 application shows how to use a **ListView** that is backed by an **ArrayAdapter**. The array that supplies values to the **ArrayAdapter** is a string array defined in the **strings.xml** file. Listing 8.1 shows the **strings.xml** file.

Listing 8.1: The res/values/strings.xml file for ListViewDemo1

```xml
<?xml version="1.0" encoding="utf-8"?>
<resources>
    <string name="app_name">ListViewDemo1</string>
    <string name="action_settings">Settings</string>

    <string-array name="players">
        <item>Player 1</item>
        <item>Player 2</item>
        <item>Player 3</item>
        <item>Player 4</item>
        <item>Player 5</item>
    </string-array>
</resources>
```

The layout for the **ArrayAdapter** is defined in the **list_item.xml** file presented in Listing 8.2. It is located under **res/layout** and contains a **TextView** element. This layout will be used as the layout for each item in the **ListView**.

Listing 8.2: The list_item.xml file

```xml
<?xml version="1.0" encoding="utf-8"?>
<TextView xmlns:android="http://schemas.android.com/apk/res/android"
    android:id="@+id/list_item"
        android:layout_width="fill_parent"
        android:layout_height="fill_parent"
        android:padding="7dip"
        android:textSize="16sp"
        android:textColor="@android:color/holo_green_dark"
        android:textStyle="bold" >
</TextView>
```

The application consists of only one activity, **MainActivity**. The layout file (**activity_main.xml**) for the activity is given in Listing 8.3 and the **MainActivity** class in Listing 8.4.

Listing 8.3: The activity_main.xml file for ListViewDemo1

```xml
<LinearLayout
    xmlns:android="http://schemas.android.com/apk/res/android"
    android:orientation="vertical"
    android:layout_width="fill_parent"
    android:layout_height="fill_parent">
    <ListView
        android:id="@+id/listView1"
        android:layout_width="wrap_content"
        android:layout_height="wrap_content" />
</LinearLayout>
```

Listing 8.4: The MainActivity class for ListViewDemo1

```java
package com.example.listviewdemo1;
import android.app.Activity;
import android.app.AlertDialog;
import android.os.Bundle;
import android.util.Log;
```

```java
import android.view.Menu;
import android.view.View;
import android.widget.AdapterView;
import android.widget.ArrayAdapter;
import android.widget.ListView;

public class MainActivity extends Activity {
    @Override
    protected void onCreate(Bundle savedInstanceState) {
        super.onCreate(savedInstanceState);
        setContentView(R.layout.activity_main);
        String[] values = getResources().getStringArray(
                R.array.players);

        ArrayAdapter<String> adapter = new ArrayAdapter<String>(
                this, R.layout.list_item, values);

        ListView listView = (ListView) findViewById(R.id.listView1);
        listView.setAdapter(adapter);
        listView.setOnItemClickListener(new
                AdapterView.OnItemClickListener() {
            @Override
            public void onItemClick(AdapterView<?> parent,
                    final View view, int position, long id) {
                String item = (String)
                        parent.getItemAtPosition(position);
                AlertDialog.Builder builder = new
                        AlertDialog.Builder(MainActivity.this);
                builder.setMessage("Selected item: "
                        + item).setTitle("ListView");
                builder.create().show();
                Log.d("ListView", "Selected item : " + item);
            }
        });
    }

    @Override
    public boolean onCreateOptionsMenu(Menu menu) {
        getMenuInflater().inflate(R.menu.menu_main, menu);
        return true;
    }
}
```

Figure 8.2 shows the application.

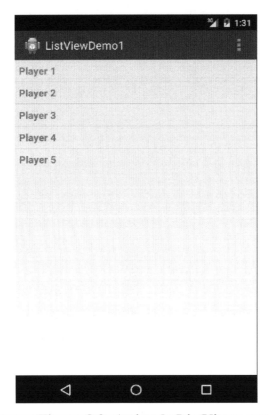

Figure 8.2: A simple ListView

Extending ListActivity and Writing A Custom Adapter

If your activity will only have one component that is a **ListView**, you should consider extending the **ListActivity** class instead of **Activity**. With **ListActivity**, you do not need a layout file for your activity. **ListActivity** already contains a **ListView** and you do not need to attach a listener to it. On top of that, the **ListActivity** class already defines a **setListAdapter** method, so you just need to call it in your **onCreate** method. In addition, instead of creating an **AdapterView.OnItemClickListener**, you just need to override the **ListActivity**'s **onListItemClick** method, which will be called when an item on the **ListView** gets selected.

The ListViewDemo2 application shows how to use **ListActivity**. The application also demonstrates how to create a custom **ListAdapter** by extending the **ArrayAdapter** class and creating a layout file for the custom **ListAdapter**.

The layout file for the custom **ListAdapter** in ListViewDemo2 is presented in Listing 8.5. It is named **pretty_adapter.xml** file and is located under **res/layout**.

Listing 8.5: The pretty_adapter.xml file

```xml
<?xml version="1.0" encoding="utf-8"?>
<LinearLayout
    xmlns:android="http://schemas.android.com/apk/res/android"
    android:layout_width="match_parent"
    android:layout_height="match_parent">

    <ImageView
        android:id="@+id/icon"
        android:layout_width="36dp"
        android:layout_height="fill_parent"/>
    <TextView
        android:id="@+id/label"
        android:layout_width="fill_parent"
        android:layout_height="fill_parent"
        android:gravity="center_vertical"
        android:padding="12dp"
        android:textSize="18sp"
        android:textColor="@android:color/holo_blue_bright"/>
</LinearLayout>
```

Listing 8.6 shows the custom adapter class, called **PrettyAdapter**.

Listing 8.6: The PrettyAdapter class

```java
package com.example.listviewdemo2;
import android.content.Context;
import android.graphics.drawable.Drawable;
import android.view.LayoutInflater;
import android.view.View;
import android.view.ViewGroup;
import android.widget.ArrayAdapter;
import android.widget.ImageView;
import android.widget.TextView;

public class PrettyAdapter extends ArrayAdapter<String> {
    private LayoutInflater inflater;
    private String[] items;
    private Drawable icon;
    private int viewResourceId;

    public PrettyAdapter(Context context,
            int viewResourceId, String[] items, Drawable icon) {
        super(context, viewResourceId, items);
        inflater = (LayoutInflater) context
                .getSystemService(Context.LAYOUT_INFLATER_SERVICE);
        this.items = items;
        this.icon = icon;
        this.viewResourceId = viewResourceId;
    }

    @Override
    public int getCount() {
```

```
            return items.length;
        }

        @Override
        public String getItem(int position) {
            return items[position];
        }

        @Override
        public long getItemId(int position) {
            return 0;
        }

        @Override
        public View getView(int position, View convertView,
                ViewGroup parent) {
            convertView = inflater.inflate(viewResourceId, null);

            ImageView imageView = (ImageView)
                    convertView.findViewById(R.id.icon);
            imageView.setImageDrawable(icon);

            TextView textView = (TextView)
                    convertView.findViewById(R.id.label);
            textView.setText(items[position]);
            return convertView;
        }
    }
}
```

The custom adapter must override several methods, notably the **getView** method, which must return a **View** that will be used for each item on the **ListView**. In this example, the view contains an **ImageView** and a **TextView**. The text for the **TextView** is taken from the array passed to the **PrettyAdapter** instance.

The last piece of the application is the **MainActivity** class in Listing 8.7. It extends **ListActivity** and is the only activity in the application.

Listing 8.7: The MainActivity class for ListViewDemo2

```
package com.example.listviewdemo2;
import android.app.ListActivity;
import android.content.Context;
import android.content.res.Resources;
import android.graphics.drawable.Drawable;
import android.os.Bundle;
import android.util.Log;
import android.view.View;
import android.widget.ListView;

public class MainActivity extends ListActivity {

    @Override
    public void onCreate(Bundle savedInstanceState) {
        super.onCreate(savedInstanceState);
```

```
        // Since we're extending ListActivity, we do
        // not need to call setContentView();

        Context context = getApplicationContext();
        Resources resources = context.getResources();

        String[] items = resources.getStringArray(
                R.array.players);
        Drawable drawable = resources.getDrawable(
                R.drawable.pretty);

        setListAdapter(new PrettyAdapter(context,
                R.layout.pretty_adapter, items, drawable));
    }

    @Override
    public void onListItemClick(ListView listView,
            View view, int position, long id) {
        Log.d("listView2", "listView:" + listView +
                ", view:" + view.getClass() +
                ", position:" + position );
    }
}
```

If you run the application, you will see an activity like that in Figure 8.3.

Figure 8.3: Using custom adapter in ListActivity

Styling the Selected Item

It is often desirable that the user be able to see clearly the currently selected item in a **ListView**. To make the selected item look differently than the rest of the items, set the **ListView**'s choice mode to **CHOICE_MODE_SINGLE**, like so.

```
listView.setChoiceMode(ListView.CHOICE_MODE_SINGLE);
```

Then, when constructing the underlying **ListAdapter**, use a layout with an appropriate style. The easiest is to pass the **simple_list_item_activated_1** field. For example, when used in a **ListView**, the following **ArrayAdapter** will cause the selected item to have a blue background.

```
ArrayAdapter<String> adapter = new ArrayAdapter<String>(
        context, android.R.layout.simple_list_item_activated_1,
        array);
```

If the default style does not appeal to you, you can create your own style by creating a selector. A selector is a drawable that can be used as a background drawable in a **TextView**. Here is an example of a selector file that must be saved in the **res/drawable** directory.

```
<selector
    xmlns:android="http://schemas.android.com/apk/res/android">
    <item android:state_activated="true"
        android:drawable="@drawable/activated"/>
</selector>
```

The selector must have an item whose **state_activated** attribute is set to **true** and whose **drawable** attribute refers to another drawable.

The ListViewDemo3 application contains an activity that employs two **ListView**s that are placed side by side. The first **ListView** on the left is given the default style whereas the second **ListView** is decorated using a custom style. Figure 8.4 shows the application.

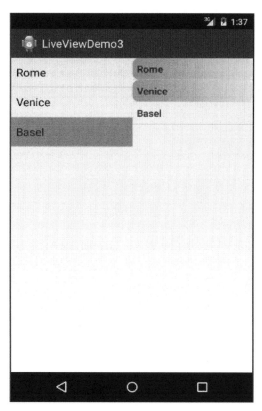

Figure 8.4: Styling the selected item of a ListView

Now, let's look at the code.

Let's start with the activity layout file in Listing 8.8.

Listing 8.8: The layout file for the main activity (activity_main.xml)

```xml
<LinearLayout
    xmlns:android="http://schemas.android.com/apk/res/android"
    android:layout_width="match_parent"
    android:layout_height="match_parent"
    android:orientation="horizontal">

    <ListView
        android:id="@+id/listView1"
        android:layout_weight="1"
        android:layout_width="0dp"
        android:layout_height="match_parent"/>
    <ListView
        android:id="@+id/listView2"
        android:layout_weight="1"
        android:layout_width="0dp"
        android:layout_height="match_parent"/>
</LinearLayout>
```

The layout uses a horizontal **LinearLayout** that contains two **ListView**s, named

listView1 and **listView2**, respectively. Both **ListView**s receive the same value for their **layout_weight** attribute, so they will have the same width when rendered.

The **MainActivity** class in Listing 8.9 represents the activity for the application. It's **onCreate** method loads both **ListView**s and pass them a **ListAdapter**. In additon, the first **ListView**'s choice mode is set to **CHOICE_MODE_SINGLE**, making a single item selectable at a time. The second **ListView**'s choice mode is set to **CHOICE_MODE_MULTIPLE**, which makes multiple items selectable at a time.

Listing 8.9: The MainActivity class

```
package com.example.listviewdemo3;
import android.app.Activity;
import android.os.Bundle;
import android.widget.ArrayAdapter;
import android.widget.ListView;
public class MainActivity extends Activity {

    @Override
    protected void onCreate(Bundle savedInstanceState) {
        super.onCreate(savedInstanceState);
        setContentView(R.layout.activity_main);
        String[] cities = {"Rome", "Venice", "Basel"};
        ArrayAdapter<String> adapter1 = new
                ArrayAdapter<String>(this,
                android.R.layout.simple_list_item_activated_1,
                cities);
        ListView listView1 = (ListView)
                findViewById(R.id.listView1);
        listView1.setAdapter(adapter1);
        listView1.setChoiceMode(ListView.CHOICE_MODE_SINGLE);

        ArrayAdapter<String> adapter2 = new
                ArrayAdapter<String>(this,
                R.layout.list_item, cities);
        ListView listView2 = (ListView)
                findViewById(R.id.listView2);
        listView2.setAdapter(adapter2);
        listView2.setChoiceMode(ListView.CHOICE_MODE_MULTIPLE);
    }
}
```

The first **ListView**'s **ListAdapter** is given the default layout (**simple_list_item_activated_1**). The second **ListView**'s **ListAdapter**, on the other hand, is set to use a layout that is pointed by **R.layout.list_item**. This refers to the **res/layout/list_item.xml** file shown in Listing 8.10.

Listing 8.10: The list_item.xml file

```
<?xml version="1.0" encoding="utf-8"?>
<TextView xmlns:android="http://schemas.android.com/apk/res/android"
    android:id="@+id/list_item"
    android:layout_width="fill_parent"
    android:layout_height="fill_parent"
    android:padding="7dip"
    android:textSize="16sp"
```

```
        android:textStyle="bold"
        android:background="@drawable/list_selector"
/>
```

A layout file for a **ListView** must contain a **TextView**, as the **list_item.xml** file does. Note that its **background** attribute is given the value **drawable/list_selector**, which references the **list_selector.xml** file in Listing 8.11. This is a selector file that will be used to style the selected item on **listView2**. The **selector** element contains an item whose **state_activated** attribute is set to **true**, which means it will be used to style the selected item. Its **drawable** attribute is set to **drawable/activated**, referring to the **drawable/activated.xml** file in Listing 8.12.

Listing 8.11: The drawable/list_selector.xml file

```
<?xml version="1.0" encoding="utf-8"?>
<selector
    xmlns:android="http://schemas.android.com/apk/res/android">
    <item android:state_activated="true"
        android:drawable="@drawable/activated"/>
</selector>
```

Listing 8.12: The drawable/activated.xml file

```
<shape xmlns:android="http://schemas.android.com/apk/res/android"
    android:shape="rectangle">
    <corners android:radius="8dp"/>
    <gradient
        android:startColor="#FFFF0000"
        android:endColor="#FFFF00"
        android:angle="45"/>
</shape>
```

The drawable in Listing 8.12 is based on an XML file that contains a shape with a given gradient color.

Running the application will give you an activity like that in Figure 8.4.

Summary

A **ListView** is a view that contains a list of scrollable items and gets its data source and layout from a **ListAdapter**, which in turn can be created from an **ArrayAdapter**. In this chapter you learned how to use the **ListView**. You also learned how to use the **ListActivity** and style the selected item on a **ListView**.

Chapter 9
GridView

A **GridView** is a view that can display a list of scrollable items in a grid. It is like a **ListVIew** except that it display items in multiple columns, unlike a **ListView** where items are displayed in a single column. Like a **ListView**, a **GridView** too takes its data source and layout from a **ListAdapter**.

This chapter shows how you can use the **GridView** widget and presents a sample application. You should have read Chapter 8, "ListView" before reading this chapter.

Overview

The **android.widget.GridView** class is the template for creating a **GridView**. Both the **GridView** and **ListView** classes are direct descendants of **android.view.AbsListView**. Like a **ListView**, a **GridView** gets its data source from a **ListAdapter**. Please refer to Chapter 8, "ListView" for more information on the **ListAdapter**.

You can use a **GridView** just like you would other views: by declaring a node in a layout file. In the case of a **GridView**, you would use this **GridView** element:

```
<GridView
    android:id="@+id/gridView1"
    android:layout_width="fill_parent"
    android:layout_height="fill_parent"
    android:columnWidth="120dp"
    android:numColumns="auto_fit"
    android:verticalSpacing="10dp"
    android:horizontalSpacing="10dp"
    android:stretchMode="columnWidth"
/>
```

You can then find the **GridView** in your activity class using **findViewById** and pass a **ListAdapter** to it.

```
GridView gridView = (GridView) findViewById(R.id.gridView1);
gridView.setAdapter(listAdapter);
```

Optionally, you can pass an **AdapterView.OnItemClickListener** to a **GridView**'s **setOnItemClickListener** method to respond to item selection:

```
gridview.setOnItemClickListener(
        new AdapterView.OnItemClickListener() {
    public void onItemClick(AdapterView<?> parent, View v, int
```

```
        position, long id) {

    // do something here


    }
});
```

Using the GridView

The GridViewDemo1 application shows you how to use the **GridView**. The application only has an activity, which uses a **GridView** to fill its entire display area. The **GridView** in turn uses a custom **ListAdapter** for its items and layout.

Listing 9.1 shows the application manifest.

Listing 9.1: The AndroidManifest.xml file

```xml
<?xml version="1.0" encoding="utf-8"?>
<manifest xmlns:android="http://schemas.android.com/apk/res/android"
    package="com.example.gridviewdemo1"
    android:versionCode="1"
    android:versionName="1.0" >

    <uses-sdk
        android:minSdkVersion="18"
        android:targetSdkVersion="18" />

    <application
        android:allowBackup="true"
        android:icon="@drawable/ic_launcher"
        android:label="@string/app_name"
        android:theme="@style/AppTheme" >
        <activity
            android:name="com.example.gridviewdemo1.MainActivity"
            android:label="@string/app_name">
            <intent-filter>
                <action android:name="android.intent.action.MAIN"/>
                <category
android:name="android.intent.category.LAUNCHER"/>
            </intent-filter>
        </activity>
    </application>
</manifest>
```

The custom **ListAdapter** that feeds the **GridView** is an instance of **GridViewAdapter**, which is presented in Listing 9.2. **GridViewAdapter** extends **android.widget.BaseAdapter**, which in turn implements the **android.widget.ListAdapter** interface. Therefore, a **GridViewAdapter** is a **ListAdapter** and can be passed to a **GridView**'s **setAdapter** method.

Listing 9.2: The GridViewAdapter class

```
package com.example.gridviewdemo1;
import android.content.Context;
import android.view.View;
import android.view.ViewGroup;
import android.widget.BaseAdapter;
import android.widget.GridView;
import android.widget.ImageView;

public class GridViewAdapter extends BaseAdapter {
    private Context context;

    public GridViewAdapter(Context context) {
        this.context = context;
    }
    private int[] icons = {
            android.R.drawable.btn_star_big_off,
            android.R.drawable.btn_star_big_on,
            android.R.drawable.alert_light_frame,
            android.R.drawable.alert_dark_frame,
            android.R.drawable.arrow_down_float,
            android.R.drawable.gallery_thumb,
            android.R.drawable.ic_dialog_map,
            android.R.drawable.ic_popup_disk_full,
            android.R.drawable.star_big_on,
            android.R.drawable.star_big_off,
            android.R.drawable.star_big_on
    };

    @Override
    public int getCount() {
        return icons.length;
    }

    @Override
    public Object getItem(int position) {
        return null;
    }

    @Override
    public long getItemId(int position) {
        return 0;
    }

    @Override
    public View getView(int position, View convertView, ViewGroup parent)
      {
        ImageView imageView;
        if (convertView == null) {
            imageView = new ImageView(context);
            imageView.setLayoutParams(new GridView.LayoutParams(100,
    100));
```

```
        imageView.setScaleType(ImageView.ScaleType.CENTER_CROP);
        imageView.setPadding(10, 10, 10, 10);
    } else {
        imageView = (ImageView) convertView;
    }
    imageView.setImageResource(icons[position]);
    return imageView;
    }
}
```

GridViewAdapter provides an implementation of the **getView** method that returns an **ImageView** displaying one of Android's default drawables:

```
private int[] icons = {
        android.R.drawable.btn_star_big_off,
        android.R.drawable.btn_star_big_on,
        android.R.drawable.alert_light_frame,
        android.R.drawable.alert_dark_frame,
        android.R.drawable.arrow_down_float,
        android.R.drawable.gallery_thumb,
        android.R.drawable.ic_dialog_map,
        android.R.drawable.ic_popup_disk_full,
        android.R.drawable.star_big_on,
        android.R.drawable.star_big_off,
        android.R.drawable.star_big_on
};
```

Now that you know what **GridViewAdapter** does, you can focus on the activity. The layout file for the activity is printed in Listing 9.3. It only consists of one component, a **GridView**.

Listing 9.3: The activity_main.xml file

```
<?xml version="1.0" encoding="utf-8"?>
<GridView xmlns:android="http://schemas.android.com/apk/res/android"
    android:id="@+id/gridview"
    android:layout_width="fill_parent"
    android:layout_height="fill_parent"
    android:columnWidth="90dp"
    android:numColumns="auto_fit"
    android:verticalSpacing="10dp"
    android:horizontalSpacing="10dp"
    android:stretchMode="columnWidth"
    android:gravity="center"
/>
```

Listing 9.4 shows the **MainActivity** class.

Listing 9.4: The MainActivity class

```
package com.example.gridviewdemo1;
import android.app.Activity;
import android.os.Bundle;
import android.view.Menu;
import android.view.View;
import android.widget.AdapterView;
```

```
import android.widget.AdapterView.OnItemClickListener;
import android.widget.GridView;
import android.widget.Toast;

public class MainActivity extends Activity {

    @Override
    protected void onCreate(Bundle savedInstanceState) {
        super.onCreate(savedInstanceState);
        setContentView(R.layout.activity_main);

        GridView gridview = (GridView) findViewById(R.id.gridview);
        gridview.setAdapter(new GridViewAdapter(this));

        gridview.setOnItemClickListener(new OnItemClickListener() {
            public void onItemClick(AdapterView<?> parent,
                    View view, int position, long id) {
                Toast.makeText(MainActivity.this, "" + position,
                    Toast.LENGTH_SHORT).show();
            }
        });
    }

    @Override
    public boolean onCreateOptionsMenu(Menu menu) {
        getMenuInflater().inflate(R.menu.menu_main, menu);
        return true;
    }
}
```

MainActivity is a simple class, with the bulk of its brain resides in its **onCreate** method. Here it loads the **GridView** from the layout and passes an instance of **GridViewAdapter** to the **GridView**'s **setAdapter** method. It also creates an **OnItemClickListener** for the **GridView** so that every time an item on the **GridView** is selected, the **onItemClick** method in the listener gets called. In this case, **onItemClick** simply creates a **Toast** that shows the position of the selected item.

Running GridViewDemo1 gives you an activity that looks like the one in Figure 9.1.

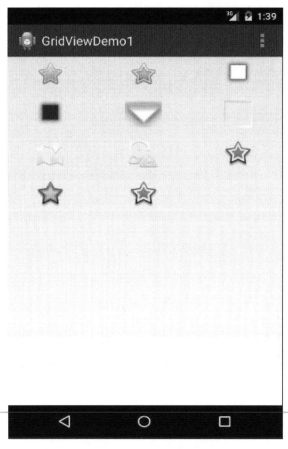

Figure 9.1: Using a GridView

Summary

A **GridView** is a view that contains a list of scrollable items displayed in a grid. Like a **ListView**, a **GridView** gets its data and layout from a **ListAdapter**. In addition, a **GridView** can also receive an **AdapterView.OnItemClickListener** to handle item selection.

Chapter 10
Styles and Themes

The look and feel of an application are governed by the styles and themes it is using. This chapter discusses these two important topics and shows you how to use them.

Overview

A view declaration in a layout file can have attributes, many of which are style-related, including **textColor**, **textSize**, **background**, and **textAppearance**.

Style-related attributes for an application can be lumped in a group and the group can be given a name and moved to a **styles.xml** file. A **styles.xml** file that is saved in the **res/values** directory will be recognized by the application as a styles file and the styles in the file can be used to style the views in the application. To apply a style to a view, use the **style** attribute. The advantage of creating a style is to make the style reusable and shareable. Styles support inheritance so you can extend a style to create a new style. Here is an example of a style in a **styles.xml** file.

```
<style name="Style1">
    <item name="android:layout_width">wrap_content</item>
    <item name="android:layout_height">wrap_content</item>
    <item name="android:textColor">#FFFFFF</item>
    <item name="android:textStyle">bold</item>
    <item name="android:textSize">25sp</item>
</style>
```

To apply the style to a view, assign the style name to the **style** attribute.

```
<TextView
    android:id="@+id/textView1"
    style="@style/Style1"
    android:text="Style 1"/>
```

Note that the **style** attribute, unlike other attributes, does not use the **android** prefix. So, it's **style** and not **android:style**.

The **TextView** element declaration above is equivalent to the following.

```
<TextView
    android:id="@+id/textView1"
    android:layout_width="wrap_content"
    android:layout_height="wrap_content"
    android:textColor="#FFFFFF"
```

```
    android:textStyle="bold"
    android:textSize="25sp"
    android:text="Style 1"/>
```

The following is a style that extends another style.

```
<style name="Style2" parent="Style1">
    <item name="android:background">
        @android:color/holo_green_light
    </item>
</style>
```

The system provides a vast collection of styles you can use in your applications. You can find a reference of all available styles in the **android.R.style** class. To use the styles listed in this class in your layout file, replace all underscores in the style name with a period. For example, you can apply the **Holo_ButtonBar** style with **@android:style/Holo.ButtonBar**.

```
<Button
    style="@android:style/Holo.ButtonBar"
    android:text="@string/hello_world"/>
```

Prefixing the value of the **style** attribute with **android** indicates that you are using a system style.

A copy of the system **styles.xml** file can be viewed here:

```
https://android.googlesource.com/platform/frameworks/base/+/refs/
heads/master/core/res/res/values/styles.xml
```

Using Styles

The StyleDemo1 application shows how you can create your own styles.

Listing 10.1 shows the application's **styles.xml** file in the **res/values** directory.

Listing 10.1: The styles.xml file

```
<resources
    xmlns:android="http://schemas.android.com/apk/res/android">
    <!-- Base application theme, dependent on API level. This theme
        is replaced by AppBaseTheme from res/values-vXX/styles.xml
        on newer devices.
    -->
    <style name="AppBaseTheme" parent="android:Theme.Light">
        <!-- Theme customizations available in newer API levels can
            go in res/values-vXX/styles.xml, while customizations
            related to backward-compatibility can go here.
        -->
    </style>
```

```
<!-- Application theme. -->
<style name="AppTheme" parent="AppBaseTheme">
    <!-- All customizations that are NOT specific to a
         particular API-level can go here. -->
</style>

<style name="WhiteOnRed">
    <item name="android:layout_width">wrap_content</item>
    <item name="android:layout_height">wrap_content</item>
    <item name="android:textColor">#FFFFFF</item>
    <item name="android:background">
        @android:color/holo_red_light
    </item>
    <item name="android:typeface">serif</item>
    <item name="android:textStyle">bold</item>
    <item name="android:textSize">25sp</item>
    <item name="android:padding">30dp</item>
</style>
<style name="WhiteOnRed.Italic">
    <item name="android:textStyle">bold|italic</item>
</style>
<style name="WhiteOnGreen" parent="WhiteOnRed">
    <item name="android:background">
        @android:color/holo_green_light
    </item>
</style>
</resources>
```

There are five styles defined in the **styles.xml** file in Listing 10.1. The first two are added
by Android Studio when the application was created. They will be explained in the
"Themes" section later in this chapter.

The other three styles are used by the main activity of the application in the layout file
for that activity. The layout file is shown in Listing 10.2.

Listing 10.2: the activity_main.xml layout file

```
<RelativeLayout
    xmlns:android="http://schemas.android.com/apk/res/android"
    xmlns:tools="http://schemas.android.com/tools"
    android:layout_width="match_parent"
    android:layout_height="match_parent"
    android:paddingBottom="@dimen/activity_vertical_margin"
    android:paddingLeft="@dimen/activity_horizontal_margin"
    android:paddingRight="@dimen/activity_horizontal_margin"
    android:paddingTop="@dimen/activity_vertical_margin"
    tools:context=".MainActivity" >

    <TextView
        android:id="@+id/textView1"
        style="@style/WhiteOnRed"
        android:text="Style WhiteOnRed" />
    <TextView
        android:id="@+id/textView2"
```

```
        android:layout_below="@id/textView1"
        android:layout_marginLeft="20sp"
        android:layout_marginTop="10sp"
        style="@style/WhiteOnRed.Italic"
        android:text="Style WhiteOnRed.Italic" />
    <TextView
        android:id="@+id/textView3"
        android:layout_below="@id/textView2"
        android:layout_toEndOf="@id/textView2"
        style="@style/WhiteOnGreen"
        android:text="Style WhiteOnGreen" />

    <TextView
        android:id="@+id/textView4"
        android:text="Style TextAppearance.Holo.Medium.Inverse"
        android:layout_below="@id/textView2"
        android:layout_width="wrap_content"
        android:layout_height="wrap_content"
        style="@android:style/TextAppearance.Holo.Medium"/>
</RelativeLayout>
```

Listing 10.3 shows the activity that uses the layout file in Listing 10.2.

Listing 10.3: The MainActivity class

```
package com.example.styledemo1;
import android.os.Bundle;
import android.app.Activity;
import android.view.Menu;

public class MainActivity extends Activity {

    @Override
    protected void onCreate(Bundle savedInstanceState) {
        super.onCreate(savedInstanceState);
        setContentView(R.layout.activity_main);
    }

    @Override
    public boolean onCreateOptionsMenu(Menu menu) {
        // Inflate the menu; this adds items to the action bar if it
        // is present.
        getMenuInflater().inflate(R.menu.menu_main, menu);
        return true;
    }

}
```

Figure 10.1 shows the StyleDemo1 application.

Figure 10.1: Using styles

Using Themes

A theme is a style that is applied to an activity or all the activities in an application. To apply a theme to an activity, use the **android:theme** attribute in the **activity** element in the manifest file. For example, the following **activity** element uses the **Theme.Holo.Light** theme.

```
<activity
    android:name="..."
    android:theme="@android:style/Theme.Holo.Light">
</activity>
```

To apply a theme to the whole application, add the **android:theme** attribute in the **application** element in the Android manifest file. For instance,

```
<application
    android:icon="@drawable/ic_launcher"
    android:label="@string/app_name"
    android:theme="@android:style/Theme.Black.NoTitleBar">

    ...
```

```
</application>
```

Android provides a collection of themes you can use in your application. A copy of the theme file can be found here:

```
https://android.googlesource.com/platform/frameworks/base/+/refs/
heads/master/core/res/res/values/themes.xml
```

Figure 10.2 to 10.4 show some of the themes that comes with Android.

Figure 10.2: Theme.Holo.Dialog.NoActionBar

Figure 10.3: Theme.Light

Figure 10.4: Theme.Holo.Light.DarkActionBar

Summary

A style is a collection of attributes that directly affect the appearance of a view. You can apply a style to a view by using the **style** attribute in the view's declaration in a layout file. A theme is a style that is applied to an activity or the entire application.

Chapter 11
Bitmap Processing

With the Android Bitmap API you can manipulate images in JPG, PNG or GIF format, such as by changing the color or the opacity of each pixel in the image. In addition, you can use the API to down-sample a large image to save memory. As such, knowing how to use this API is useful, even when you are not writing a photo editor or an image processing application.

This chapter explains how to work with bitmaps and provides an example.

Overview

A bitmap is an image file format that can store digital images independently of the display device. A bitmap simply means a map of bits. Today the term also includes other formats that support lossy and lossless compression, such as JPEG, GIF and PNG. GIF and PNG support transparency and lossless compression, whereas JPEG support lossy compression and does not support transparency. Another way of representing digital images is through mathematical expressions. Such images are known as vector graphics.

The Android framework provides an API for processing bitmap images. This API takes the form of classes, interfaces, and enums in the **android.graphics** package and its subpackages. The **Bitmap** class models a bitmap image. A **Bitmap** can be displayed on an activity using the **ImageView** widget.

The easiest way to load a bitmap is by using the **BitmapFactory** class. This class provides static methods for constructing a **Bitmap** from a file, a byte array, an Android resource or an **InputStream**. Here are some of the methods.

```
public static Bitmap decodeByteArray(byte[] data, int offset,
        int length)

public static Bitmap decodeFile(java.lang.String pathName)

public static Bitmap decodeResource(
        android.content.res.Resources res, int id)

public static Bitmap decodeStream (java.io.InputStream is)
```

For example, to construct a **Bitmap** from an Android resource in an activity class, you would use this code.

```
Bitmap bmp = BitmapFactory.decodeResource(getResources(),
        R.drawable.image1);
```

Here, **getResources** is a method in the **android.content.Context** class that returns the application's resources (**Context** is the parent class of **Activity**). The identifier (**R.drawable.image1**) allows Android to pick the correct image from the resources.

The **BitmapFactory** class also offers static methods that take options as a **BitmapFactory.Options** object:

```
public static Bitmap decodeByteArray (byte[] data, int offset,
        int length, BitmapFactory.Options opts)

public static Bitmap decodeFile (java.lang.String pathName,
        BitmapFactory.Options opts)

public static Bitmap decodeResource (android.content.res.Resources
        res, int id, BitmapFactory.Options opts)

public static Bitmap decodeStream (java.io.InputStream is,
        Rect outPadding, BitmapFactory.Options opts)
```

There are two things you can do with a **BitmapFactory.Options**. The first is it allows you to configure the resulting bitmap as the class allows you to down-sample the bitmap, set the bitmap to be mutable and configure its density. The second is you can use the **BitmapFactory.Options** to read the properties of a bitmap without actually loading the image. For example, you may pass a **BitmapFactory.Options** to one of the **decode** methods in **BitmapFactory** and read the size of the image. If the size is considered too large, then you can down-sample it, saving precious memory. Down-sampling makes sense for large bitmaps when it does not reduce render quality. For instance, a 20,000 x 10,000 bitmap can be down-sampled to 2,000 x 1,000 without degradation assuming the device screen resolution does not exceed 2,000 x 1,000. In the process, it saves a lot of memory.

To decode a **Bitmap** without actually loading the bitmap, set the **inJustDecodeBounds** field of the **BitmapFactory.Options** object to **true**.

```
BitmapFactory.Options opts = new BitmapFactory.Options()
opts.inJustDecodeBounds = true;
```

If you pass the options to one of the **decode** methods in **BitmapFactory**, the method will return null and simply populate the **BitmapFactory.Options** object that you passed. From this object, you can retrieve the bitmap size and other properties:

```
int imageHeight = options.outHeight;
int imageWidth = options.outWidth;
String imageType = options.outMimeType;
```

The **inSampleSize** field of **BitmapFactor.Options** tells the system how to sample a bitmap. A value greater than 1 indicates that the image should be down-sampled. For example, setting the **inSampleSize** field to 4 returns an image whose size is a quarter that of the original image.

Regarding this field, the Android documentation says that the decoder uses a final value based on powers of 2, which means you should only assign a power of 2, such as 2, 4, 8, and so on. However, my own test shows that this only applies to images in JPG format and does not apply to PNGs. For instance, if the width of a PNG image is 1200, assigning 3 to this field returns an image with a width of 400 pixels, which means the **inSampleSize** value does not have to be a power of two.

Finally, once you get a **Bitmap** from a **BitmapFactory**, you can pass the **Bitmap** to an **ImageView** to be displayed:

```
ImageView imageView1 = (ImageView) findViewById(...);
imageView1.setImageBitmap(bitmap);
```

Bitmap Processing

The BitmapDemo application showcases an activity that shows an **ImageView** that displays a **Bitmap** that can be down-sampled. There are four bitmaps (two JPEGs, one GIF, and one PNG) included and the application provides a button to change bitmaps. The main (and only) activity of the application is shown in Figure 11.1.

Listing 11.1 shows the **AndroidManifest.xml** file for the application.

Listing 11.1: The AndroidManifest.xml file

```xml
<?xml version="1.0" encoding="utf-8"?>
<manifest xmlns:android="http://schemas.android.com/apk/res/android"
    package="com.example.bitmapdemo"
    android:versionCode="1"
    android:versionName="1.0" >
    <uses-sdk
        android:minSdkVersion="18"
        android:targetSdkVersion="18" />
    <application
        android:allowBackup="true"
        android:icon="@drawable/ic_launcher"
        android:label="@string/app_name"
        android:theme="@style/AppTheme" >
        <activity
            android:name="com.example.bitmapdemo.MainActivity"
            android:label="@string/app_name" >
            <intent-filter>
                <action android:name="android.intent.action.MAIN"/>
                <category
android:name="android.intent.category.LAUNCHER"/>
            </intent-filter>
        </activity>
    </application>
</manifest>
```

Figure 11.1: The BitmapDemo application

There is only one activity in this application. The layout file for the activity is given in Listing 11.2.

Listing 11.2: The activity_main.xml file

```
<LinearLayout xmlns:android="http://schemas.android.com/apk/res/android"
    xmlns:tools="http://schemas.android.com/tools"
    android:layout_width="match_parent"
    android:layout_height="match_parent"
    android:orientation="vertical"
    android:gravity="bottom"
    tools:context=".MainActivity" >

    <ImageView
        android:id="@+id/image_view1"
        android:layout_width="wrap_content"
        android:layout_height="wrap_content"
        android:contentDescription="@string/text_content_desc"/>

    <LinearLayout
        android:layout_width="match_parent"
        android:layout_height="wrap_content"
```

```
        android:orientation="horizontal" >

        <TextView
            android:layout_width="wrap_content"
            android:layout_height="wrap_content"
            android:text="@string/text_sample_size"/>
        <TextView
            android:id="@+id/sample_size"
            android:layout_width="wrap_content"
            android:layout_height="wrap_content"
          />
        <Button
            android:onClick="scaleUp"
            android:layout_width="wrap_content"
            android:layout_height="wrap_content"
            android:text="@string/action_scale_up" />

        <Button
            android:onClick="scaleDown"
            android:layout_width="wrap_content"
            android:layout_height="wrap_content"
            android:text="@string/action_scale_down" />

    </LinearLayout>
    <LinearLayout
        android:layout_width="match_parent"
        android:layout_height="wrap_content"
        android:orientation="horizontal" >

        <Button
            android:onClick="changeImage"
            android:layout_width="wrap_content"
            android:layout_height="wrap_content"
            android:text="@string/action_change_image" />
        <TextView
            android:id="@+id/image_info"
            android:layout_width="wrap_content"
            android:layout_height="wrap_content"/>
    </LinearLayout>
</LinearLayout>
```

The layout contains a **LinearLayout** that in turn contains an **ImageView** and two **LinearLayout**s. The first inner layout contains two **TextView**s and buttons for scaling up and down the bitmap. The second inner layout contains a **TextView** to display the bitmap metadata and a button to change the bitmap.

The **MainActivity** class is presented in Listing 11.3.

Listing 11.3: The MainActivity class

```
package com.example.bitmapdemo;
import android.app.Activity;
import android.graphics.Bitmap;
import android.graphics.BitmapFactory;
```

```java
import android.os.Bundle;
import android.view.Menu;
import android.view.View;
import android.widget.ImageView;
import android.widget.TextView;

public class MainActivity extends Activity {
    int sampleSize = 2;
    int imageId = 1;

    @Override
    protected void onCreate(Bundle savedInstanceState) {
        super.onCreate(savedInstanceState);
        setContentView(R.layout.activity_main);
        refreshImage();
    }

    @Override
    public boolean onCreateOptionsMenu(Menu menu) {
        getMenuInflater().inflate(R.menu.menu_main, menu);
        return true;
    }

    public void scaleDown(View view) {
        if (sampleSize < 8) {
            sampleSize++;
            refreshImage();
        }
    }

    public void scaleUp(View view) {
        if (sampleSize > 2) {
            sampleSize--;
            refreshImage();
        }
    }
    private void refreshImage() {
        BitmapFactory.Options options = new BitmapFactory.Options();
        options.inJustDecodeBounds = true;
        BitmapFactory.decodeResource(getResources(),
                R.drawable.image1, options);
        int imageHeight = options.outHeight;
        int imageWidth = options.outWidth;
        String imageType = options.outMimeType;

        StringBuilder imageInfo = new StringBuilder();

        int id = R.drawable.image1;
        if (imageId == 2) {
            id = R.drawable.image2;
            imageInfo.append("Image 2.");
        } else if (imageId == 3) {
            id = R.drawable.image3;
```

```
            imageInfo.append("Image 3.");
        } else if (imageId == 4) {
            id = R.drawable.image4;
            imageInfo.append("Image 4.");
        } else {
            imageInfo.append("Image 1.");
        }
        imageInfo.append(" Original Dimension: " + imageWidth
                + " x " + imageHeight);
        imageInfo.append(". MIME type: " + imageType);
        options.inSampleSize = sampleSize;
        options.inJustDecodeBounds = false;
        Bitmap bitmap1 = BitmapFactory.decodeResource(
                getResources(), id, options);
        ImageView imageView1 = (ImageView)
                findViewById(R.id.image_view1);
        imageView1.setImageBitmap(bitmap1);

        TextView sampleSizeText = (TextView)
                findViewById(R.id.sample_size);
        sampleSizeText.setText("" + sampleSize);
        TextView infoText = (TextView)
                findViewById(R.id.image_info);
        infoText.setText(imageInfo.toString());

    }

    public void changeImage(View view) {
        if (imageId < 4) {
            imageId++;
        } else {
            imageId = 1;
        }
        refreshImage();
    }
}
```

The **scaleDown**, **scaleUp** and **changeImage** methods are connected to the three buttons. All methods eventually call the **refreshImage** method.

The **refreshImage** method uses the **BitmapFactory.decodeResource** method to first read the properties of the bitmap resource, by passing a **BitmapFactory.Options** whose **inJustDecodeBounds** field is set to **true**. Recall that this is a strategy for avoiding loading a large image that will take much if not all of the available memory.

```
        BitmapFactory.Options options = new BitmapFactory.Options();
        options.inJustDecodeBounds = true;
        BitmapFactory.decodeResource(getResources(),
                R.drawable.image1, options);
```

It then reads the dimension and image type of the bitmap.

```
        int imageHeight = options.outHeight;
        int imageWidth = options.outWidth;
```

```
String imageType = options.outMimeType;
```

Next, it sets the **inJustDecodeBounds** field to **false** and uses the **sampleSize** value (that the user can change by clicking the Scale Up or Scale Down button) to set the **inSampleSize** field of the **BitmapFactory.Options**, and decode the bitmap for the second time.

```
options.inSampleSize = sampleSize;
options.inJustDecodeBounds = false;
Bitmap bitmap1 = BitmapFactory.decodeResource(
        getResources(), id, options);
```

The dimension of the resulting Bitmap will be determined by the value of the **inSampleSize** field.

Summary

The Android Bitmap API centers around the **BitmapFactory** and **Bitmap** classes. The former provides static methods for constructing a **Bitmap** object from an Android resource, a file, an InputStream, or a byte array. Some of the methods can take a **BitmapFactory.Options** to determine what kind of bitmap they will produce. The resulting bitmap can then be assigned to an **ImageView** for display.

Chapter 12
Graphics and Custom Views

Thanks to Android's extensive library, you have dozens of views and widgets at your disposal. If none of these meets your need, you can create a custom view and draw directly on it using the Android Graphics API.

This chapter discusses the use of some members of the Graphics API to draw on a canvas and create a custom view. A sample application called **CanvasDemo** is presented at the end of this chapter.

Overview

The Android Graphics API comprises the members of the **android.graphics** package. The **Canvas** class in this package plays a central role in 2D graphics. You can get an instance of **Canvas** from the system and you do not need to create one yourself. Once you have an instance of **Canvas**, you can call its various methods, such as **drawColor**, **drawArc**, **drawRect**, **drawCircle**, and **drawText**.

In addition to **Canvas**, **Color** and **Paint** are frequently used. A **Color** object represents a color code as an **int**. The **Color** class defines a number of color code fields and methods for creating and converting color **int**s. Color code fields defined in **Color** includes **BLACK**, **CYAN**, **MAGENTA**, **YELLOW**, **WHITE**, **RED**, **GREEN** and **BLUE**.

Take the **drawColor** method in **Canvas** as an example. This method accepts a color code as an argument.

```
public void drawColor(int color);
```

drawColor changes the color of the canvas with the specified color. To change the canvas color to magenta, you would write

```
canvas.drawColor(Color.MAGENTA);
```

A **Paint** is required when drawing a shape or text. A **Paint** determines the color and transparency of the shape or text drawn as well as the font family and style of the text.

To create a **Paint**, use one of the **Paint** class's constructors:

```
public Paint()
public Paint(int flags)
public Paint(Paint anotherPaint)
```

If you use the second constructor, you can pass one or more fields defined in the **Paint** class. For example, the following code creates a **Paint** by passing the **LINEAR_TEXT_FLAG** and **ANTI_ALIAS_FLAG** fields.

```
Paint paint = new Paint(
        Paint.LINEAR_TEXT_FLAG | Paint.ANTI_ALIAS_FLAG);
```

Hardware Acceleration

Modern smart phones and tablets come with a graphic processing unit (GPU), an electronic circuit that specializes in image creation and rendering. Starting with Android 3.0, the Android framework will utilize any GPU it can find on a device, resulting in improved performance through hardware acceleration. Hardware acceleration is enabled by default for any application targeting Android API level 14 or above.

Unfortunately, currently not all drawing operations work when hardware acceleration is turned on. You can disable hardware acceleration by setting the **android:hardwareAccelerated** attribute to **false** in either the **application** or **activity** element in your android manifest file. For example, to turn off hardware acceleration for the whole application, use this:

```
<application android:hardwareAccelerated="false">
```

To disable hardware acceleration in an activity, use this:

```
<activity android:hardwareAccelerated="false" />
```

It is possible to use the **android:hardwareAccelerated** attribute in both application or activity levels. For example, the following indicates that all except one activity in the application should use hardware acceleration.

```
<application android:hardwareAccelerated="true">
    <activity ... />
    <activity android:hardwareAccelerated="false" />
</application>
```

Note
To try out the examples in this chapter, you must disable hardware acceleration.

Creating A Custom View

To create a custom view, extend the **android.view.View** class or one of its subclasses and override its **onDraw** method. Here is the signature of **onDraw**.

```
protected void onDraw (android.graphics.Canvas canvas)
```

The system calls the **onDraw** method and pass a **Canvas**. You can use the methods in **Canvas** to draw shapes and text. You can also create path and regions to draw more complex shapes.

The **onDraw** method may be called many times during the application lifecycle. As such, you should not perform expensive operations here, such as allocating objects. Objects that you need to use in **onDraw** should be created somewhere else.

For example, most drawing methods in **Canvas** require a **Paint**. Rather than creating a **Paint** in **onDraw**, you should create it at the class level and have it available for use in **onDraw**. This is illustrated in the following class.

```java
public class MyCustomView extends View {
    Paint paint;
    {
        paint = ... // create a Paint object here
    }
    @Override
    protected void onDraw(Canvas canvas) {
        // use paint here.
    }
}
```

Drawing Basic Shapes

The **Canvas** class defines methods such as **drawLine**, **drawCircle**, and **drawRect** to draw shapes. For example, the following code shows how you can draw shapes in your **onDraw** method.

```java
Paint paint = new Paint(Paint.FAKE_BOLD_TEXT_FLAG);

protected void onDraw(Canvas canvas) {
    // change canvas background color.
    canvas.drawColor(Color.parseColor("#bababa"));

    // draw basic shapes
    canvas.drawLine(5,   5, 200,  5, paint);
    canvas.drawLine(5,  15, 200, 15, paint);
    canvas.drawLine(5,  25, 200, 25, paint);

    paint.setColor(Color.YELLOW);
    canvas.drawCircle(50, 70, 35, paint);

    paint.setColor(Color.GREEN);
    canvas.drawRect(new Rect(100, 60, 150, 80), paint);

    paint.setColor(Color.DKGRAY);
    canvas.drawOval(new RectF(160, 60, 250, 80), paint);

    ...
}
```

Figure 12.1 shows the result.

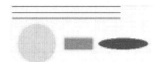

Figure 12.1: Basic shapes

Drawing Text

To draw text on a canvas, use the **drawText** method and a **Paint**. For example, the following code draws text using different colors.

```
// draw text
textPaint.setTextSize(22);
canvas.drawText("Welcome", 20, 100, textPaint);
textPaint.setColor(Color.MAGENTA);
textPaint.setTextSize(40);
canvas.drawText("Welcome", 20, 140, textPaint);
```

Figure 12.2 shows the drawn text.

Figure 12.2: Drawing text

Transparency

Android's Graphics API supports transparency. You can set the transparency by assigning an alpha value to the **Paint** used in drawing. Consider the following code.

```
// transparency
textPaint.setColor(0xFF465574);
textPaint.setTextSize(60);
canvas.drawText("Android Rocks", 20, 340, textPaint);
// opaque circle
canvas.drawCircle(80, 300, 20, paint);
// semi-transparent circles
paint.setAlpha(110);
canvas.drawCircle(160, 300, 39, paint);
paint.setColor(Color.YELLOW);
paint.setAlpha(140);
canvas.drawCircle(240, 330, 30, paint);
paint.setColor(Color.MAGENTA);
paint.setAlpha(30);
canvas.drawCircle(288, 350, 30, paint);
paint.setColor(Color.CYAN);
paint.setAlpha(100);
canvas.drawCircle(380, 330, 50, paint);
```

Figure 12.3 shows some semi transparent circles..

Figure 12.3: Transparency

Shaders

A **Shader** is a span of colors. You create a **Shader** by defining two colors as in the following code.

```
// shader
Paint shaderPaint = new Paint();
Shader shader = new LinearGradient(0, 400, 300, 500, Color.RED,
        Color.GREEN, Shader.TileMode.CLAMP);
shaderPaint.setShader(shader);
canvas.drawRect(0, 400, 200, 500, shaderPaint);
```

Figure 12.4 shows a linear gradient shader.

Figure 12.4: Using a linear gradient shader

Clipping

Clipping is the process of allocating an area on a canvas for drawing. The clipped area can be a rectangle, a circle, or any arbitrary shape you can imagine. Once you clip the canvas, any other drawing that would otherwise be rendered outside the area will be ignored.

Figure 12.5 shows a clip area in the shape of a star. After the canvas is clipped, drawn text will only be visible within the clipped area.

Figure 12.5: An example of clipping

The **Canvas** class provides the following methods for clipping: **clipRect**, **clipPath**, and **clipRegion**. The **clipRect** method uses a **Rect** as a clip area and **clipPath** uses a **Path**. For example, the clip area in Figure 12.5 was created using this code.

```
canvas.clipPath(starPath);
// starPath is a Path in the shape of a star, see next section
// on how to create it.
textPaint.setColor(Color.parseColor("yellow"));
canvas.drawText("Android", 350, 550, textPaint);
textPaint.setColor(Color.parseColor("#abde97"));
canvas.drawText("Android", 400, 600, textPaint);
canvas.drawText("Android Rocks", 300, 650, textPaint);
canvas.drawText("Android Rocks", 320, 700, textPaint);
canvas.drawText("Android Rocks", 360, 750, textPaint);
canvas.drawText("Android Rocks", 320, 800, textPaint);
```

You'll learn more about clipping in the next sections.

Using Paths

A **Path** is a collection of any number of straight line segments, quadratic curves, and cubic curves. A **Path** can be used for clipping or to draw text on.

As an example, this method creates a star path. It takes a coordinate that is the location of its center.

```
private Path createStarPath(int x, int y) {
    Path path = new Path();
    path.moveTo(0 + x, 150 + y);
    path.lineTo(120 + x, 140 + y);
    path.lineTo(150 + x, 0 + y);
    path.lineTo(180 + x, 140 + y);
    path.lineTo(300 + x, 150 + y);
    path.lineTo(200 + x, 190 + y);
    path.lineTo(250 + x, 300 + y);
    path.lineTo(150 + x, 220 + y);
```

```
        path.lineTo(50 + x, 300 + y);
        path.lineTo(100 + x, 190 + y);
        path.lineTo(0 + x, 150 + y);
        return path;
    }
```

The following code shows how to draw text that curves along a **Path**.

```
public class CustomView extends View {
    Path curvePath;
    Paint textPaint = new Paint(Paint.LINEAR_TEXT_FLAG);
    {
        Typeface typeface = Typeface.create(Typeface.SERIF,
                Typeface.BOLD);
        textPaint.setTypeface(typeface);
        curvePath = createCurvePath();
    }

    private Path createCurvePath() {
        Path path = new Path();
        path.addArc(new RectF(400, 40, 780, 300), -210, 230);
        return path;
    }

    protected void onDraw(Canvas canvas) {
        ...
        // draw text on path
        textPaint.setColor(Color.rgb(155, 20, 10));
        canvas.drawTextOnPath("Nice artistic touches",
                curvePath, 10, 10, textPaint);
        ...
    }
}
```

Figure 12.6 shows the drawn text.

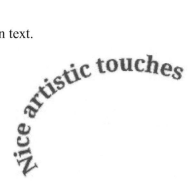

Figure 12.6: Drawing text on a path

The CanvasDemo Application

The CanvasDemo application features a custom view and contains all the code snippets presented in this chapter. Figure 12.7 shows the main activity of the application.

Figure 12.7: The CanvasDemo application

Listing 12.1 shows the **AndroidManifest.xml** file for this application. It only has one activity.

Listing 12.1: The AndroidManifest.xml file

```
<?xml version="1.0" encoding="utf-8"?>
<manifest xmlns:android="http://schemas.android.com/apk/res/android"
    package="com.example.canvasdemo"
    android:versionCode="1"
    android:versionName="1.0" >

    <uses-sdk
        android:minSdkVersion="18"
```

```
            android:targetSdkVersion="18" />

    <application
        android:hardwareAccelerated="false"
        android:allowBackup="true"
        android:icon="@drawable/ic_launcher"
        android:label="@string/app_name"
        android:theme="@style/AppTheme" >
        <activity
            android:name="com.example.canvasdemo.MainActivity"
            android:label="@string/app_name" >
            <intent-filter>
                <action android:name="android.intent.action.MAIN"/>
                <category
android:name="android.intent.category.LAUNCHER" />
            </intent-filter>
        </activity>
    </application>
</manifest>
```

The main actor of the application is the **CustomView** class in Listing 12.2. It extends **View** and overrides its **onDraw** method.

Listing 12.2: The CustomView class

```
package com.example.canvasdemo;
import android.content.Context;
import android.graphics.Canvas;
import android.graphics.Color;
import android.graphics.LinearGradient;
import android.graphics.Paint;
import android.graphics.Path;
import android.graphics.Rect;
import android.graphics.RectF;
import android.graphics.Shader;
import android.graphics.Typeface;
import android.view.View;

public class CustomView extends View {

    public CustomView(Context context) {
        super(context);
    }
    Paint paint = new Paint(Paint.FAKE_BOLD_TEXT_FLAG);
    Path starPath;
    Path curvePath;


    Paint textPaint = new Paint(Paint.LINEAR_TEXT_FLAG);
    Paint shaderPaint = new Paint();
    {
        Typeface typeface = Typeface.create(
                Typeface.SERIF, Typeface.BOLD);
        textPaint.setTypeface(typeface);
```

```
        Shader shader = new LinearGradient(0, 400, 300, 500,
                Color.RED, Color.GREEN, Shader.TileMode.CLAMP);
        shaderPaint.setShader(shader);
        // create star path
        starPath = createStarPath(300, 500);
        curvePath = createCurvePath();
    }

    protected void onDraw(Canvas canvas) {
        // draw basic shapes
        canvas.drawLine(5,   5, 200,  5, paint);
        canvas.drawLine(5,  15, 200, 15, paint);
        canvas.drawLine(5,  25, 200, 25, paint);

        paint.setColor(Color.YELLOW);
        canvas.drawCircle(50, 70, 35, paint);

        paint.setColor(Color.GREEN);
        canvas.drawRect(new Rect(100, 60, 150, 80), paint);

        paint.setColor(Color.DKGRAY);
        canvas.drawOval(new RectF(160, 60, 250, 80), paint);


        // draw text
        textPaint.setTextSize(22);
        canvas.drawText("Welcome", 20, 150, textPaint);
        textPaint.setColor(Color.MAGENTA);
        textPaint.setTextSize(40);
        canvas.drawText("Welcome", 20, 190, textPaint);

        // transparency
        textPaint.setColor(0xFF465574);
        textPaint.setTextSize(60);
        canvas.drawText("Android Rocks", 20, 340, textPaint);
        // opaque circle
        canvas.drawCircle(80, 300, 20, paint);
        // semi-transparent circle
        paint.setAlpha(110);
        canvas.drawCircle(160, 300, 39, paint);
        paint.setColor(Color.YELLOW);
        paint.setAlpha(140);
        canvas.drawCircle(240, 330, 30, paint);
        paint.setColor(Color.MAGENTA);
        paint.setAlpha(30);
        canvas.drawCircle(288, 350, 30, paint);
        paint.setColor(Color.CYAN);
        paint.setAlpha(100);
        canvas.drawCircle(380, 330, 50, paint);

        // draw text on path
        textPaint.setColor(Color.rgb(155, 20, 10));
        canvas.drawTextOnPath("Nice artistic touches",
```

```
                    curvePath, 10, 10, textPaint);


        // shader
        canvas.drawRect(0, 400, 200, 500, shaderPaint);


        // create a star-shaped clip
        canvas.drawPath(starPath, textPaint);
        textPaint.setColor(Color.CYAN);
        canvas.clipPath(starPath);
        textPaint.setColor(Color.parseColor("yellow"));
        canvas.drawText("Android", 350, 550, textPaint);
        textPaint.setColor(Color.parseColor("#abde97"));
        canvas.drawText("Android", 400, 600, textPaint);
        canvas.drawText("Android Rocks", 300, 650, textPaint);
        canvas.drawText("Android Rocks", 320, 700, textPaint);
        canvas.drawText("Android Rocks", 360, 750, textPaint);
        canvas.drawText("Android Rocks", 320, 800, textPaint);
    }

    private Path createStarPath(int x, int y) {
        Path path = new Path();
        path.moveTo(0 + x, 150 + y);
        path.lineTo(120 + x, 140 + y);
        path.lineTo(150 + x, 0 + y);
        path.lineTo(180 + x, 140 + y);
        path.lineTo(300 + x, 150 + y);
        path.lineTo(200 + x, 190 + y);
        path.lineTo(250 + x, 300 + y);
        path.lineTo(150 + x, 220 + y);
        path.lineTo(50 + x, 300 + y);
        path.lineTo(100 + x, 190 + y);
        path.lineTo(0 + x, 150 + y);
        return path;
    }

    private Path createCurvePath() {
        Path path = new Path();
        path.addArc(new RectF(400, 40, 780, 300),
                -210, 230);
        return path;
    }
}
```

The **MainActivity** class, given in Listing 12.3, instantiates the **CustomView** class and pass the instance to its **setContentView** method. This is unlike most applications in this book where you pass a layout resource identifier to another overload of **setContentView**.

Listing 12.3: The MainActivity class

```
package com.example.canvasdemo;
import android.app.Activity;
import android.os.Bundle;
```

```
public class MainActivity extends Activity {
    @Override
    protected void onCreate(Bundle savedInstanceState) {
        super.onCreate(savedInstanceState);
        CustomView customView = new CustomView(this);
        setContentView(customView);
    }
}
```

Summary

The Android SDK comes with a wide range of views that you can use in your applications. If none of these suits your need, you can create a custom view and draw on it. This chapter showed you how to create a custom view and draw multiple shapes on a canvas.

Chapter 13
Fragments

A powerful feature added to Android 3.0 (API level 11), fragments are components that can be embedded into an activity. Unlike custom views, fragments have their own lifecycle and may or may not have a user interface.

This chapter explains what fragments are and shows how to use them.

The Fragment Lifecycle

You create a fragment by extending the **android.app.Fragment** class or one of its subclasses. A fragment may or may not have a user interface. A fragment with no user interface (UI) normally acts as a worker for the activity the fragment is embedded into. If a fragment has a UI, it may contain views arranged in a layout file that will be loaded after the fragment is created. In many aspects, writing a fragment is similar to writing an activity.

In order to create fragments effectively, you need to know the lifecycle of a fragment. Figure 13.1 shows the lifecycle of a fragment.

The lifecycle of a fragment is similar to that of an activity. For example, it has callback methods such as **onCreate**, **onResume** and **onPause**. On top of that, there are additional methods like **onAttach**, **onActivityCreated** and **onDetach**. **onAttach** is called after the fragment is associated with an activity and **onActivityCreated** gets called after the **onCreate** method of the activity that contains the fragment is completed. **onDetach** is invoked before a fragment is detached from an activity.

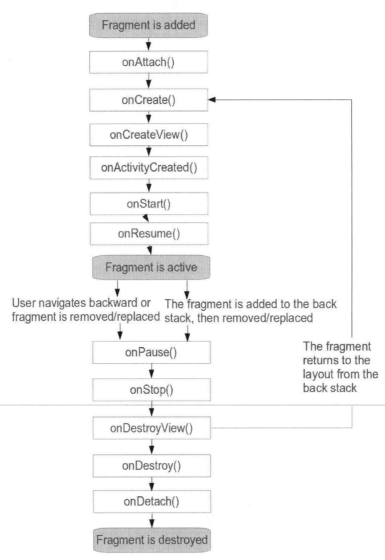

Figure 13.1: The fragment lifecycle

- **onAttach**. Called right after the fragment is associated with its activity.
- **onCreate**. Called to create the fragment the first time.
- **onCreateView**. Called when it is time to create the layout for the fragment. It must return the fragment's root view.
- **onActivityCreated**. Called to tell the fragment that its activity's **onCreate** method has completed.
- **onStart**. Called when the fragment's view is made visible to the user.
- **onResume**. Called when the containing activity enters the **Resumed** state, which means the activity is running.
- **onPause**. Called when the containing activity is being paused.
- **onStop**. Called when the containing activity is stopped.
- **onDestroyView**. Called to allow the fragment to release resources used for its

view.
- **onDestroy**. Called to allow the fragment to do final clean-up before it is destroyed.
- **onDetach**. Called right after the fragment is detached from its activity.

There are some subtle differences between an activity and a fragment. In an activity, you normally set the view for the activity in its **onCreate** method using the **setContentView** method, e.g.

```
protected void onCreate(android.os.Bundle savedInstanceState) {
    super(savedInstanceState);
    setContentView(R.layout.activity_main);
    ...
}
```

In a fragment you normally create a view in its **onCreateView** method. Here is the signature of the **onCreateView** method.

```
public View onCreateView(android.view.LayoutInflater inflater,
        android.view.ViewGroup container,
        android.os.Bundle savedInstanceState);
```

Noticed there are three arguments that are passed to **onCreateView**? The first argument is a **LayoutInflater** that you use to inflate any view in the fragment. The second argument is the parent view the fragment should be attached to. The third argument, a **Bundle**, if not null contains information from the previously saved state.

In an activity, you can obtain a reference to a view by calling the **findViewById** method on the activity. In a fragment, you can find a view in the fragment by calling the **findViewById** on the parent view.

```
View root = inflater.inflate(R.layout.fragment_names,
        container, false);
View aView = (View) root.findViewById(id);
```

Also note that a fragment should not know anything about its activity or other fragments. If you need to listen for an event that occurs in a fragment that affects the activity or other views or fragments, do not write a listener in the fragment class. Instead, trigger a new event in response to the fragment event and let the activity handle it.

You will learn more about creating a fragment in later sections in this chapter.

Fragment Management

To use a fragment in an activity, use the **fragment** element in a layout file just as you would a view. Specify the fragment class name in the **android:name** attribute and an identifier in the **android:id** attribute. Here is an example of a **fragment** element.

```
<fragment
    android:name="com.example.MyFragment"
    android:id="@+id/fragment1"
    ...
/>
```

Alternatively, You can manage fragments programmatically in your activity class using an **android.app.FragmentManager**. You can obtain the default instance of **FragmentManager** by calling the **getFragmentManager** method in your activity class. Then, call the **beginTransaction** method on the **FragmentManager** to obtain a **FragmentTransaction**.

```
FragmentManager fragmentManager = getFragmentManager();
FragmentTransaction fragmentTransaction =
        fragmentManager.beginTransaction();
```

The **android.app.FragmentTransaction** class offers methods for adding, removing, and replacing fragments. Once you're finished, call **FragmentTransaction.commit()** to commit your changes.

You can add a fragment to an activity using one of the **add** method overloads in the **FragmentTransaction** class. You have to specify to which view the fragment should be added to. Normally, you would add a fragment to a **FrameLayout** or some other type of layout. Here is one of the **add** methods in **FragmentTransaction**.

```
public abstract FragmentTransaction add(int containerViewId,
        Fragment fragment, String tag)
```

To use **add**, you would instantiate your fragment class and then specify the ID of the view to add to. If you pass a tag, you can later retrieve your fragment using the **findFragmentByTag** method on the **FragmentManager**.

If you are not using a tag, you can use this **add** method.

```
public abstract FragmentTransaction add(int containerViewId,
        Fragment fragment)
```

To remove a fragment from an activity, call the **remove** method on the **FragmentTransaction**.

```
public abstract FragmentTransaction remove(Fragment fragment)
```

And to replace a fragment in a view with another fragment, use the **replace** method.

```
public abstract FragmentTransaction replace(int containerViewId,
        Fragment fragment, String tag)
```

As a last step once you are finished managing your fragments, call commit on the **FragmentTransaction**.

```
public abstract int commit()
```

Using A Fragment

The FragmentDemo1 application is a sample application with an activity that uses two fragments. The first fragment lists some cities. Selecting a city causes the second fragment to show the picture of the selected city. Since proper design dictates that a fragment should not know anything about its surrounding, the first fragment rises an event upon receiving user selection. The activity handles this new event and causes the second fragment to change.

Figure 13.2 shows how FragmentDemo1 looks like.

Figure 13.2: Using fragments

The manifest for the application is printed in Listing 13.1.

Listing 13.1: The AndroidManifest.xml file for FragmentDemo1

```
<?xml version="1.0" encoding="utf-8"?>
<manifest xmlns:android="http://schemas.android.com/apk/res/android"
    package="com.example.fragmentdemo1"
    android:versionCode="1"
    android:versionName="1.0" >

    <uses-sdk
        android:minSdkVersion="18"
        android:targetSdkVersion="18" />

    <application
        android:allowBackup="true"
        android:icon="@drawable/ic_launcher"
        android:label="@string/app_name"
        android:theme="@style/AppTheme">
        <activity
            android:name="com.example.fragmentdemo1.MainActivity"
            android:label="@string/app_name" >
            <intent-filter>
```

```
            <action android:name="android.intent.action.MAIN"/>
            <category
    android:name="android.intent.category.LAUNCHER"/>
        </intent-filter>
      </activity>
    </application>
</manifest>
```

Nothing extraordinary here. It simply declares an activity for the application.

You use a fragment as you would a view or a widget, by declaring it in an activity's layout file or by programmatically adding one. For FragmentDemo1, two fragments are added to the layout of the application's main activity. The layout file is shown in Listing 13.2.

Listing 13.2: The layout file for the main activity (activity_main.xml)

```
<LinearLayout
    xmlns:android="http://schemas.android.com/apk/res/android"
    android:orientation="horizontal"
    android:layout_width="match_parent"
    android:layout_height="match_parent">
    <fragment
        android:name="com.example.fragmentdemo1.NamesFragment"
        android:id="@+id/namesFragment"
        android:layout_weight="1"
        android:layout_width="0dp"
        android:layout_height="match_parent" />
    <fragment
        android:name="com.example.fragmentdemo1.DetailsFragment"
        android:id="@+id/detailsFragment"
        android:layout_weight="2.5"
        android:layout_width="0dp"
        android:layout_height="match_parent" />
</LinearLayout>
```

The layout for the main activity uses a horizontal **LinearLayout** that splits the screen into two panes. The ratio of the pane widths is 1:2.5, as defined by the **layout_weight** attributes of the **fragment** elements. Each pane is filled with a fragment. The first pane is represented by the **NamesFragment** class and the second pane by the **DetailsFragment** class.

The first fragment, **NamesFragment**, gets its layout from the **fragment_names.xml** file in Listing 13.3. This file is located in the **res/layout** folder.

Listing 13.3: The fragment_names.xml file

```
<ListView
    xmlns:android="http://schemas.android.com/apk/res/android"
    android:id="@+id/listView1"
    android:layout_width="wrap_content"
    android:layout_height="wrap_content"
    android:background="#FFFF55"/>
```

The layout of **NamesFragment** is very simple. It contains a single view that is a **ListView**. The layout is loaded in the **onCreateView** method of the fragment class (See

Listing 13.4).

Listing 13.4: The NamesFragment class

```java
package com.example.fragmentdemo1;
import android.app.Activity;
import android.app.Fragment;
import android.os.Bundle;
import android.view.LayoutInflater;
import android.view.View;
import android.view.ViewGroup;
import android.widget.AdapterView;
import android.widget.ArrayAdapter;
import android.widget.ListView;

public class NamesFragment extends Fragment {
    @Override
    public View onCreateView(LayoutInflater inflater,
            ViewGroup container, Bundle savedInstanceState) {
        final String[] names = {"Amsterdam", "Brussels", "Paris"};
        // use android.R.layout.simple_list_item_activated_1
        // to have the selected item in a different color
        ArrayAdapter<String> adapter = new ArrayAdapter<String>(
                getActivity(),
                android.R.layout.simple_list_item_activated_1,
                names);

        View view = inflater.inflate(R.layout.fragment_names,
                container, false);
        final ListView listView = (ListView) view.findViewById(
                R.id.listView1);

        listView.setChoiceMode(ListView.CHOICE_MODE_SINGLE);
        listView.setOnItemClickListener(new
                AdapterView.OnItemClickListener() {
            @Override
            public void onItemClick(AdapterView<?> parent,
                    final View view, int position, long id) {
                if (callback != null) {
                    callback.onItemSelected(names[position]);
                }
            }
        });
        listView.setAdapter(adapter);
        return view;
    }

    public interface Callback {
        public void onItemSelected(String id);
    }

    private Callback callback;

    @Override
```

```
    public void onAttach(Activity activity) {
        super.onAttach(activity);
        if (activity instanceof Callback) {
            callback = (Callback) activity;
        }
    }
    @Override
    public void onDetach() {
        super.onDetach();
        callback = null;
    }
}
```

The **NamesFragment** class defines a **Callback** interface that its activity must implement to listen to the item selection event of its **ListView**. The activity can then use it to drive the second fragment. The **onAttach** method makes sure that the implementing class is an **Activity**.

The second fragment, **DetailsFragment**, has a layout file that is given in Listing 13.5. It contains a **TextView** and an **ImageView**. The **TextView** displays the name of the selected city and the **ImageView** shows the picture of the selected city.

Listing 13.5: The fragment_details.xml file

```xml
<LinearLayout
    xmlns:android="http://schemas.android.com/apk/res/android"
    android:orientation="vertical"
    android:background="#FAFAD2"
    android:layout_width="match_parent"
    android:layout_height="match_parent">
    <TextView
        android:id="@+id/text1"
        android:layout_width="wrap_content"
        android:layout_height="wrap_content"
        android:textSize="30sp"/>
    <ImageView
        android:id="@+id/imageView1"
        android:layout_width="match_parent"
        android:layout_height="match_parent"/>
</LinearLayout>
```

The **DetailsFragment** class in shown in Listing 13.6. It has a **showDetails** method that the containing activity can call to change the content of the **TextView** and **ImageView**.

Listing 13.6: The DetailsFragment class

```java
package com.example.fragmentdemo1;
import android.app.Fragment;
import android.os.Bundle;
import android.view.LayoutInflater;
import android.view.View;
import android.view.ViewGroup;
import android.widget.ImageView;
import android.widget.ImageView.ScaleType;
import android.widget.TextView;
```

```java
public class DetailsFragment extends Fragment {

    @Override
    public View onCreateView(LayoutInflater inflater,
            ViewGroup container, Bundle savedInstanceState) {
        return inflater.inflate(R.layout.fragment_details,
                container, false);
    }

    public void showDetails(String name) {
        TextView textView = (TextView)
                getView().findViewById(R.id.text1);
        textView.setText(name);

        ImageView imageView = (ImageView) getView().findViewById(
                R.id.imageView1);
        imageView.setScaleType(ScaleType.FIT_XY); // stretch image
        if (name.equals("Amsterdam")) {
            imageView.setImageResource(R.drawable.amsterdam);
        } else if (name.equals("Brussels")) {
            imageView.setImageResource(R.drawable.brussels);
        } else if (name.equals("Paris")) {
            imageView.setImageResource(R.drawable.paris);
        }
    }
}
```

The activity class for FragmentDemo1 is presented in Listing 13.7.

Listing 13.7: The activity class for FragmentDemo1

```java
package com.example.fragmentdemo1;
import android.app.Activity;
import android.os.Bundle;

public class MainActivity extends Activity
        implements NamesFragment.Callback {

    @Override
    protected void onCreate(Bundle savedInstanceState) {
        super.onCreate(savedInstanceState);
        setContentView(R.layout.activity_main);
    }
    @Override
    public void onItemSelected(String value) {
        DetailsFragment details = (DetailsFragment)
                getFragmentManager().findFragmentById(
                        R.id.detailsFragment);
        details.showDetails(value);
    }
}
```

The most important thing to note is that the activity class implements
NamesFragment.Callback so that it can capture the item click event in the fragment. The

onItemSelected method is an implementation for the **Callback** interface. It calls the **showDetails** method in the second fragment to change the text and picture of the selected city.

Extending ListFragment and Using FragmentManager

FragmentDemo1 showed how you could add a fragment to an activity using the **fragment** element in the activity's layout file. In the second sample application, FragmentDemo2, you will learn how to add a fragment to an activity programmatically.

FragmentDemo2 is similar in functionality to its predecessor with a few differences. The first difference pertains to how the name and the picture of a selected city are updated. In FragmentDemo1, the containing activity calls the **showDetails** method in the second fragment, passing the city name. In FragmentDemo2, when a city is selected, the activity creates a new instance of **DetailsFragment** and uses it to replace the old instance.

The second difference is the fact that the first fragment extends **ListFragment** instead of **Fragment**. **ListFragment** is a subclass of **Fragment** and contains a **ListView** that fills its entire view. When subclassing **ListFragment**, you should override its **onCreate** method and call its **setListAdapter** method. This is demonstrated in the **NamesListFragment** class in Listing 13.8.

Listing 13.8: The NamesListFragment class

```
package com.example.fragmentdemo2;
import android.app.Activity;
import android.app.ListFragment;
import android.os.Bundle;
import android.view.View;
import android.widget.AdapterView;
import android.widget.ArrayAdapter;
import android.widget.ListView;

/* we don't need fragment_names-xml anymore */
public class NamesListFragment extends ListFragment {

    final String[] names = {"Amsterdam", "Brussels", "Paris"};

    @Override
    public void onCreate(Bundle savedInstanceState) {
        super.onCreate(savedInstanceState);
        ArrayAdapter<String> adapter = new ArrayAdapter<String>(
                getActivity(),
                android.R.layout.simple_list_item_activated_1,
                names);
        setListAdapter(adapter);
    }

    @Override
    public void onViewCreated(View view,
```

```
                Bundle savedInstanceState) {
        // ListView can only be accessed here, not in onCreate()
        super.onViewCreated(view, savedInstanceState);
        ListView listView = getListView();
        listView.setChoiceMode(ListView.CHOICE_MODE_SINGLE);
        listView.setOnItemClickListener(new
                AdapterView.OnItemClickListener() {
            @Override
            public void onItemClick(AdapterView<?> parent,
                    final View view, int position, long id) {
                if (callback != null) {
                    callback.onItemSelected(names[position]);
                }
            }
        });

    }

    public interface Callback {
        public void onItemSelected(String id);
    }

    private Callback callback;

    @Override
    public void onAttach(Activity activity) {
        super.onAttach(activity);
        if (activity instanceof Callback) {
            callback = (Callback) activity;
        }
    }
    @Override
    public void onDetach() {
        super.onDetach();
        callback = null;
    }
}
```

Like the **NamesFragment** class in FragmentDemo1, the **NamesListFragment** class in
FragmentDemo2 also defines a **Callback** interface that a containing activity must
implement to listen to the **ListView**'s **OnItemClick** event.

The second fragment, **DetailsFragment** in Listing 13.9, expects its activity to pass
two arguments, a name and an image ID. In its **onCreate** method, the fragment retrieves
these arguments and store them in class level variables, **name** and **imageId**. The values of
the variables are then used in its **onCreateView** method to populate its **TextView** and
ImageView.

Listing 13.9: The DetailsFragment class

```
package com.example.fragmentdemo2;
import android.app.Fragment;
import android.os.Bundle;
import android.view.LayoutInflater;
```

```
import android.view.View;
import android.view.ViewGroup;
import android.widget.ImageView;
import android.widget.ImageView.ScaleType;
import android.widget.TextView;

public class DetailsFragment extends Fragment {

    int imageId;
    String name;

    public DetailsFragment() {
    }

    @Override
    public void onCreate(Bundle savedInstanceState) {
        super.onCreate(savedInstanceState);
        if (getArguments().containsKey("name")) {
            name = getArguments().getString("name");
        }
        if (getArguments().containsKey("imageId")) {
            imageId = getArguments().getInt("imageId");
        }
    }

    @Override
    public View onCreateView(LayoutInflater inflater,
            ViewGroup container, Bundle savedInstanceState) {

        View rootView = inflater.inflate(
                R.layout.fragment_details, container, false);
        TextView textView = (TextView)
                rootView.findViewById(R.id.text1);
        textView.setText(name);

        ImageView imageView = (ImageView) rootView.findViewById(
                R.id.imageView1);
        imageView.setScaleType(ScaleType.FIT_XY); //stretch image
        imageView.setImageResource(imageId);
        return rootView;
    }
}
```

Now that you have looked at the fragments, take a close look at the activity. The layout file is given in Listing 13.10. Instead of two fragment elements like in FragmentDemo1, the activity layout file in FragmentDemo2 has a fragment element and a **FrameLayout**. The latter acts as the container for the second fragment.

Listing 13.10: The activity_main.xml file

```
<LinearLayout
    xmlns:android="http://schemas.android.com/apk/res/android"
    android:orientation="horizontal"
    android:layout_width="match_parent"
```

```
            android:layout_height="match_parent">
        <fragment
            android:name="com.example.fragmentdemo2.NamesListFragment"
            android:id="@+id/namesFragment"
            android:layout_weight="1"
            android:layout_width="0dp"
            android:layout_height="match_parent"/>
        <FrameLayout
            android:id="@+id/details_container"
            android:layout_width="0dp"
            android:layout_height="match_parent"
            android:layout_weight="2.5"/>
</LinearLayout>
```

The activity class for FragmentDemo2 is given in Listing 13.11. Like the activity class in FragmentDemo1, it also implements the **Callback** interface. However, its implementation of the **onItemSelected** method is different. First, it passes two arguments to the **DetailsFragment**. Second, every time **onItemSelected** is called, a new **DetailsFragment** instance is created and passed to the **FrameLayout**.

Listing 13.11: The MainActivity class

```java
package com.example.fragmentdemo2;
import android.app.Activity;
import android.app.FragmentManager;
import android.app.FragmentTransaction;
import android.os.Bundle;

public class MainActivity extends Activity
        implements NamesListFragment.Callback {

    @Override
    protected void onCreate(Bundle savedInstanceState) {
        super.onCreate(savedInstanceState);
        setContentView(R.layout.activity_main);

    }
    @Override
    public void onItemSelected(String value) {

        Bundle arguments = new Bundle();
        arguments.putString("name", value);
        if (value.equals("Amsterdam")) {
            arguments.putInt("imageId", R.drawable.amsterdam);
        } else if (value.equals("Brussels")) {
            arguments.putInt("imageId", R.drawable.brussels);
        } else if (value.equals("Paris")) {
            arguments.putInt("imageId", R.drawable.paris);
        }
        DetailsFragment fragment = new DetailsFragment();
        fragment.setArguments(arguments);
        FragmentManager fragmentManager = getFragmentManager();
        FragmentTransaction fragmentTransaction =
                fragmentManager.beginTransaction();
```

```
        fragmentTransaction.replace(
                R.id.details_container, fragment);
        fragmentTransaction.commit();
    }
}
```

Figure 13.3 shows FragmentDemo2.

Figure 13.3: FragmentDemo2

Summary

Fragments are components that can be added to an activity. A fragment has its own lifecycle and has methods that get called when certain phases of its life occur. In this chapter you have learned to write your own fragments.

Chapter 14
Multi-Pane Layouts

An Android tablet generally has a larger screen than that of a handset. In many cases, you might want to take advantage of the bigger screen in tablets to display more information by using a multi-pane layout.

This chapter discusses multi-pane layouts using fragments that you learned in Chapter 13, "Fragments."

Overview

A tablet has a larger screen than a handset and you can display more information on a tablet than on a handset. If you are writing an application that needs to look good on both types of devices, a common strategy is to support two layouts. A single-pane layout can be used for handsets and a multi-pane layout for tablets.

Figure 14.1 shows a dual-pane version of an application and Figure 14.2 shows the same application in single-pane mode.

In a single layout, you would display an activity that often contains a single fragment, which in turn often contains a **ListView**. Selecting an item on the **ListView** would start another activity.

In a multi-pane layout, you would have an activity that is big enough for two panes. You would use the same fragment, but this time when an item is selected, it updates a second fragment instead of starting another activity.

The question is, how do you tell the system to pick the right layout? Prior to Android 3.2 (API level 13) a screen may fall into one of these categories depending on its size:

- small, for screens that are at least 426dp x 320dp
- normal, for screens that are at least 470dp x 320dp
- large, for screens that are at least 640dp x 480dp
- xlarge, for screens that are at least 960dp x 720dp

Here, dp stands for density independent pixel. You can calculate the number of pixels (px) from the dp and the screen density (in dots per inch or dpi) by using this formula.

```
px = dp * (dpi / 160)
```

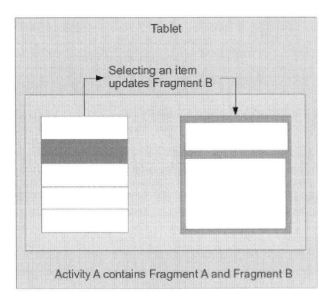

Figure 14.1: Dual-pane layout

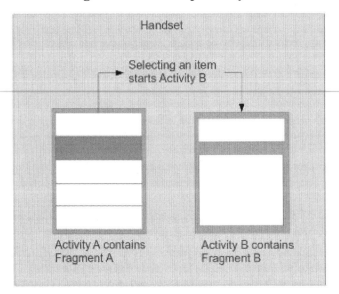

Figure 14.2: Single-pane layout

To support a screen category, you would place your layout files in the folder dedicated to that category, that is **res/layout-small** for small screens, **res/layout** for normal screens, **res/layout-large** for large screens, and **res/layout-xlarge** for xlarge screens. To support both normal and large screens, you would have layout files in both **res/layout** and **res/layout-large** directories.

The system is not without limitations, however. For example, a 7" tablet and a 10" tablet would both fall into the xlarge category, even though they provide different amounts of space. To allow for different layouts for 7" and 10" tablets, Android 3.2

changed the way it worked. Instead of the four screen sizes, Android 3.2 and later employ a new technique that measures the screen based on the amount of space in dp, rather than trying to make the layout fit the generalized size groups.

With the new system, it is easy to provide different layouts for tablets with a 600dp screen width (such as in a typical 7" tablet) and tablets with a 720dp screen width (such as in a typical 10" tablet). A typical handset, by the way, has a 320dp screen width.

Now, to support large screen devices for both pre-3.2 devices and later devices, you need to store layout files in both **res/layout-large** and **res/layout-sw600dp** directories. In other words, for each layout you end up with three files (assuming your layout file is called **main.xml**):

- **res/layout/main.xml** for normal screens
- **res/layout-large/main.xml** for devices running pre-3.2 Android having a large screen
- **res/layout-sw600dp/main.xml** for devices running Android 3.2 or later having a large screen

In addition, if your application has a different screen for 10" tablets, you will also need a **res/layout-sw720dp/main.xml** file.

The **main.xml** files in the **layout-large** and **layout-sw600dp** directories are identical and having duplicates that both have to be changed if one of them was updated is certainly a maintenance nightmare.

To get around it, you can use references. With references, you only need two layout files, one for normal screens and one for large screens, both in the **res/layout** directory. Assuming the names of your layout files are **main.xml** and **main_large.xml**, to reference the latter, you need to have a **refs.xml** file in both **res/values-large** and **res/values-sw600dp**. The content of **refs.xml** would be as follows.

```
<resources>
    <item name="main" type="layout">@layout/main_large</item>
</resources>
```

Figure 14.3 shows the content of the **res** directory.

This way, you still have two identical files, the **refs.xml** file in the **values-large** directory and the **refs.xml** file in the **values-sw600dp** directory. However, these are reference files that do not need to be updated if the layout changes.

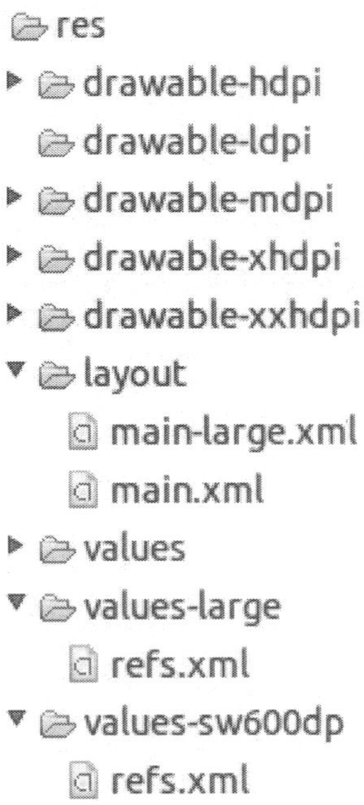

```
📂 res
  ▶ 📂 drawable-hdpi
    📂 drawable-ldpi
  ▶ 📂 drawable-mdpi
  ▶ 📂 drawable-xhdpi
  ▶ 📂 drawable-xxhdpi
  ▼ 📂 layout
      📄 main-large.xml
      📄 main.xml
  ▶ 📂 values
  ▼ 📂 values-large
      📄 refs.xml
  ▼ 📂 values-sw600dp
      📄 refs.xml
```

Figure 14.3: The structure of the res directory that supports layout references

A Multi-Pane Example

MultiPaneDemo is an application that supports small and large screens. For large screens it shows an activity that uses a multi-pane layout consisting of two fragments. For smaller screens, another activity will be shown that contains only one fragment.

The easiest way to create a multi-pane application is by using Android Studio. As usual, you would use the New Android Application wizard as described in Chapter 1, "Getting Started." However, instead of creating a blank activity as in Chapter 1, you should select Master/Detail Flow, as shown in Figure 14.4.

Figure 14.4: Selecting Master/Detail Flow activity

After you reach the window in Figure 14.4, click **Next**. In the window that appears next (See Figure 14.5), select the name for your item(s) and click **Finish**.

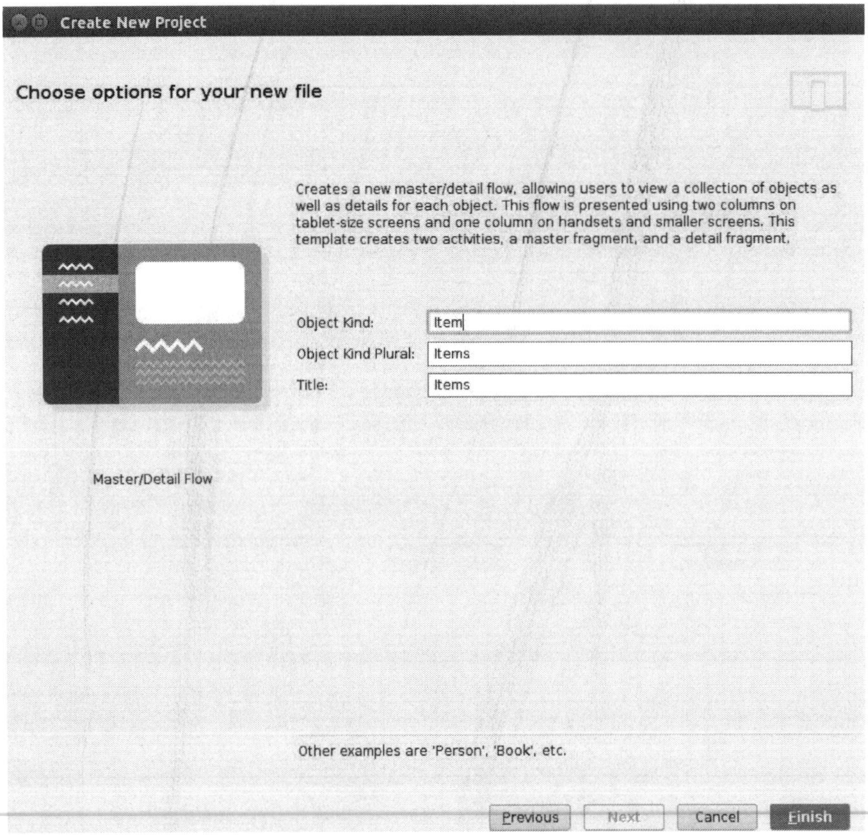

Figure 14.5: Choosing names for the items

Android Studio supports multi-pane/single-pane layouts by creating two versions of the layout file for the main activity. The single-pane version is stored in the **res/layout** directory and the multi-pane version in the res/layout-sw600dp directory. When the application is launched, the main activity automatically selects the correct layout file depending on the screen resolution.

Android Studio also creates a multi-pane application that supports Android 3.0 and later as well as pre-3.0 Android. If you don't need to support older devices, however, you can remove the support classes. The advantage is you will have an apk file that is about 30KB lighter.

The **AndroidManifest.xml** file for the application is given in Listing 14.1.

Listing 14.1: The AndroidManifest.xml file

```xml
<?xml version="1.0" encoding="utf-8"?>
<manifest xmlns:android="http://schemas.android.com/apk/res/android"
    package="com.example.multipanedemo"
    android:versionCode="1"
    android:versionName="1.0" >

    <uses-sdk
        android:minSdkVersion="18"
```

```
                    android:targetSdkVersion="18" />

    <application
        android:allowBackup="true"
        android:icon="@drawable/ic_launcher"
        android:label="@string/app_name"
        android:theme="@style/AppTheme" >
        <activity
            android:name=".ItemListActivity"
            android:label="@string/app_name" >
            <intent-filter>
                <action android:name="android.intent.action.MAIN" />
                <category
android:name="android.intent.category.LAUNCHER" />
            </intent-filter>
        </activity>
        <activity
            android:name=".ItemDetailActivity"
            android:label="@string/title_item_detail"
            android:parentActivityName=".ItemListActivity" >
            <meta-data
                android:name="android.support.PARENT_ACTIVITY"
                android:value=".ItemListActivity" />
        </activity>
    </application>

</manifest>
```

The application has two activities. The main activity is used in both single-pane and multi-pane environments. The second activity is used in the single-pane environment only.

The Layouts and Activities

As you can see in the manifest, the **ItemListActivity** class is the activity class that will be instantiated when the application is launched. This class is shown in Listing 14.2.

Listing 14.2: The ItemListActivity class

```
package com.example.multipanedemo;
import android.app.Activity;
import android.content.Intent;
import android.os.Bundle;

public class ItemListActivity extends Activity
        implements ItemListFragment.Callbacks {

    private boolean twoPane;

    @Override
    protected void onCreate(Bundle savedInstanceState) {
        super.onCreate(savedInstanceState);
        setContentView(R.layout.activity_item_list);
```

```
        if (findViewById(R.id.item_detail_container) != null) {
            twoPane = true;

            // In two-pane mode, list items should be given the
            // 'activated' state when touched.
            ((ItemListFragment) getFragmentManager()
                    .findFragmentById(R.id.item_list))
                    .setActivateOnItemClick(true);
        }
    }

    /**
     * Callback method from {@link ItemListFragment.Callbacks}
     * indicating that the item with the given ID was selected.
     */
    @Override
    public void onItemSelected(String id) {
        if (twoPane) {
            Bundle arguments = new Bundle();
            arguments.putString(ItemDetailFragment.ARG_ITEM_ID, id);
            ItemDetailFragment fragment = new ItemDetailFragment();
            fragment.setArguments(arguments);
            getFragmentManager().beginTransaction()
                    .replace(R.id.item_detail_container, fragment)
                    .commit();

        } else {
            // In single-pane mode, simply start the detail activity
            // for the selected item ID.
            Intent detailIntent = new Intent(this,
        ItemDetailActivity.class);
            detailIntent.putExtra(ItemDetailFragment.ARG_ITEM_ID, id);
            startActivity(detailIntent);
        }
    }
}
```

The **onCreate** method in **ItemListActivity** loads the layout indicated by layout identifier **R.layout.activity_item_list**.

```
    protected void onCreate(Bundle savedInstanceState) {
        super.onCreate(savedInstanceState);
        setContentView(R.layout.activity_item_list);
    ...
```

In devices with smaller screens, the **res/layout/activity_item_list.xml** will be loaded. In devices with larger screens, the system will try to locate the **activity_item_list.xml** file in either the **res/layout-large** or **res/layout-sw600dp** directory.

The multi-pane **activity_item_list.xml** file in res/layout/sw600dp is used in devices with a large screen. This layout file is presented in Listing 14.3.

Listing 14.3: The res/layout-sw600dp/activity_item_list.xml file (multi-pane)

```
<LinearLayout
    xmlns:android="http://schemas.android.com/apk/res/android"
    xmlns:tools="http://schemas.android.com/tools"
    android:layout_width="match_parent"
    android:layout_height="match_parent"
    android:layout_marginLeft="16dp"
    android:layout_marginRight="16dp"
    android:baselineAligned="false"
    android:divider="?android:attr/dividerHorizontal"
    android:orientation="horizontal"
    android:showDividers="middle"
    tools:context=".ItemListActivity">

    <!--
    This layout is a two-pane layout for the Items
    master/detail flow.

    -->

    <fragment android:id="@+id/item_list"
        android:name="com.example.multipanedemo.ItemListFragment"
        android:layout_width="0dp"
        android:layout_height="match_parent"
        android:layout_weight="1"
        tools:layout="@android:layout/list_content" />

    <FrameLayout android:id="@+id/item_detail_container"
        android:layout_width="0dp"
        android:layout_height="match_parent"
        android:layout_weight="3" />

</LinearLayout>
```

The **activity_item_list.xml** layout file features a horizontal **LinearLayout** that splits the screen into two panes. The left pane consists of a fragment that contains a **ListView**. The right pane contains a **FrameLayout** to which instances of another fragment called **ItemDetailFragment** can be added. Listing 14.4 shows the layout for **ItemDetailFragment**.

Listing 14.4: The fragment_item_detail.xml file

```
<TextView xmlns:android="http://schemas.android.com/apk/res/android"
    xmlns:tools="http://schemas.android.com/tools"
    android:id="@+id/item_detail"
    style="?android:attr/textAppearanceLarge"
    android:layout_width="match_parent"
    android:layout_height="match_parent"
    android:padding="16dp"
    android:textIsSelectable="true"
    tools:context=".ItemDetailFragment" />
```

For smaller screens, two activities will be used. The main activity will load the

activity_item_list.xml layout file in Listing 14.5. This layout contains the same fragment used by the left pane in the multi-pane layout.

Listing 14.5: The res/layout/activity_item_list.xml file (single-pane)

```
<fragment xmlns:android="http://schemas.android.com/apk/res/android"
    xmlns:tools="http://schemas.android.com/tools"
    android:id="@+id/item_list"
    android:name="com.example.multipanedemo.ItemListFragment"
    android:layout_width="match_parent"
    android:layout_height="match_parent"
    android:layout_marginLeft="16dp"
    android:layout_marginRight="16dp"
    tools:context=".ItemListActivity"
    tools:layout="@android:layout/list_content" />
```

The Fragment Classes

The two fragment classes are given in Listing 14.6 and Listing 14.7, respectively.

Listing 14.6: The ItemListFragment class

```
package com.example.multipanedemo;
import android.app.Activity;
import android.os.Bundle;
import android.app.ListFragment;
import android.view.View;
import android.widget.ArrayAdapter;
import android.widget.ListView;
import com.example.multipanedemo.dummy.DummyContent;

public class ItemListFragment extends ListFragment {

    private static final String STATE_ACTIVATED_POSITION =
        "activated_position";

    /**
     * The fragment's current callback object, which is notified of
     * list item clicks.
     */
    private Callbacks mCallbacks = sDummyCallbacks;

    /**
     * The current activated item position. Only used on tablets.
     */
    private int mActivatedPosition = ListView.INVALID_POSITION;

    /**
     * A callback interface that all activities containing this
     * fragment must implement. This mechanism allows
     * activities to be notified of item selections.
     */
    public interface Callbacks {
        /**
         * Callback for when an item has been selected.
```

```java
     */
    public void onItemSelected(String id);
}

/**
 * A dummy implementation of the {@link Callbacks} interface
 * that does nothing. Used only when this fragment is not
 * attached to an activity.
 */
private static Callbacks sDummyCallbacks = new Callbacks() {
    @Override
    public void onItemSelected(String id) {
    }
};

/**
 * Mandatory empty constructor for the fragment manager to
 * instantiate the fragment (e.g. upon screen orientation
 * changes).
 */
public ItemListFragment() {
}

@Override
public void onCreate(Bundle savedInstanceState) {
    super.onCreate(savedInstanceState);

    // TODO: replace with a real list adapter.
    setListAdapter(new ArrayAdapter<DummyContent.DummyItem>(
            getActivity(),
            android.R.layout.simple_list_item_activated_1,
            android.R.id.text1,
            DummyContent.ITEMS));
}

@Override
public void onViewCreated(View view, Bundle savedInstanceState) {
    super.onViewCreated(view, savedInstanceState);

    // Restore the previously serialized activated item
    // position.
    if (savedInstanceState != null
            && savedInstanceState.containsKey(
                STATE_ACTIVATED_POSITION)) {
        setActivatedPosition(savedInstanceState.getInt(
            STATE_ACTIVATED_POSITION));
    }
}

@Override
public void onAttach(Activity activity) {
    super.onAttach(activity);
```

```
        // Activities containing this fragment must implement its
        // callbacks.
        if (!(activity instanceof Callbacks)) {
            throw new IllegalStateException(
                "Activity must implement fragment's callbacks.");
        }

        mCallbacks = (Callbacks) activity;
    }

    @Override
    public void onDetach() {
        super.onDetach();

        // Reset the active callbacks interface to the dummy
        // implementation.
        mCallbacks = sDummyCallbacks;
    }

    @Override
    public void onListItemClick(ListView listView, View view, int
            position, long id) {
        super.onListItemClick(listView, view, position, id);

        // Notify the active callbacks interface (the activity, if
        // the fragment is attached to one) that an item has been
        // selected.
        mCallbacks.onItemSelected(DummyContent.ITEMS.get(
                position).id);
    }

    @Override
    public void onSaveInstanceState(Bundle outState) {
        super.onSaveInstanceState(outState);
        if (mActivatedPosition != ListView.INVALID_POSITION) {
            // Serialize and persist the activated item position.
            outState.putInt(STATE_ACTIVATED_POSITION,
                    mActivatedPosition);
        }
    }

    /**
     * Turns on activate-on-click mode. When this mode is on, list
     * items will be
     * given the 'activated' state when touched.
     */
    public void setActivateOnItemClick(boolean activateOnItemClick) {
        // When setting CHOICE_MODE_SINGLE, ListView will
        // automatically
        // give items the 'activated' state when touched.
        getListView().setChoiceMode(activateOnItemClick
                ? ListView.CHOICE_MODE_SINGLE
                : ListView.CHOICE_MODE_NONE);
```

```
    }

    private void setActivatedPosition(int position) {
        if (position == ListView.INVALID_POSITION) {
            getListView().setItemChecked(mActivatedPosition, false);
        } else {
            getListView().setItemChecked(position, true);
        }
        mActivatedPosition = position;
    }
}
```

The **ItemListFragment** class extends **ListFragment** and gets the data for its **ListView**
from a **DummyContent** class. It also provides a **Callbacks** interface that any activity
using this fragment must implement to handle the **ListItemClick** event of the **ListView**.
In the **onAttach** method, the fragment makes sure the activity class implements Callbacks
and replaces the content of mCallbacks with the activity, in effect delegating the event
handling to the activity.

Listing 14.7: The ItemDetailFragment class

```
package com.example.multipanedemo;
import android.os.Bundle;
import android.app.Fragment;
import android.view.LayoutInflater;
import android.view.View;
import android.view.ViewGroup;
import android.widget.TextView;
import com.example.multipanedemo.dummy.DummyContent;

/**
 * A fragment representing a single Item detail screen.
 * This fragment is either contained in a {@link ItemListActivity}
 * in two-pane mode (on tablets) or a {@link ItemDetailActivity}
 * on handsets.
 */
public class ItemDetailFragment extends Fragment {
    /**
     * The fragment argument representing the item ID that this
     * fragment represents.
     */
    public static final String ARG_ITEM_ID = "item_id";

    /**
     * The dummy content this fragment is presenting.
     */
    private DummyContent.DummyItem mItem;

    /**
     * Mandatory empty constructor for the fragment manager to
     * instantiate the fragment (e.g. upon screen orientation
     * changes).
     */
    public ItemDetailFragment() {
```

```
    }

    @Override
    public void onCreate(Bundle savedInstanceState) {
        super.onCreate(savedInstanceState);

        if (getArguments().containsKey(ARG_ITEM_ID)) {
            // Load the dummy content specified by the fragment
            // arguments. In a real-world scenario, use a Loader
            // to load content from a content provider.
            mItem = DummyContent.ITEM_MAP.get(
                    getArguments().getString(ARG_ITEM_ID));
        }
    }

    @Override
    public View onCreateView(LayoutInflater inflater, ViewGroup
            container, Bundle savedInstanceState) {
        View rootView =
                inflater.inflate(R.layout.fragment_item_detail,
                container, false);

        // Show the dummy content as text in a TextView.
        if (mItem != null) {
            ((TextView) rootView.findViewById(R.id.item_detail))
                    .setText(mItem.content);
        }

        return rootView;
    }
}
```

Running the Application

Figure 14.6 and Figure 14.7 show the MultipaneDemo1 application on a tablet and a handset, respectively.

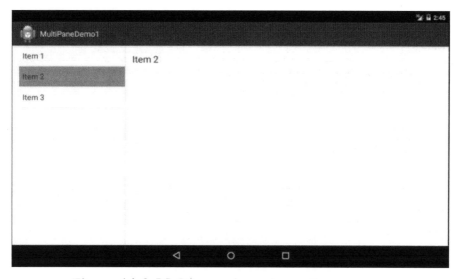

Figure 14.6: Multi-pane layout on a large screen

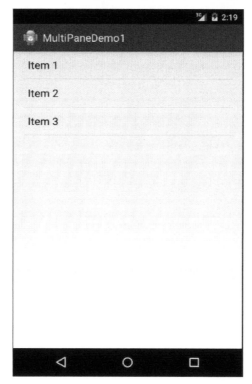

Figure 14.7: Single-pane layout on a small screen

Summary

To give your users the best experience, you may want to use different layouts for different screen sizes. In this chapter, you learned that a good strategy to achieve that is to use a multi-pane layout for tablets and a single-pane layout for handsets.

Chapter 15
Animation

Animation is an interesting feature in Android that has been available since the very beginning (API Level 1). In this chapter you will learn to use an Animation API called property animation, which was added to Honeycomb (API Level 11). The new API is more powerful than the previous animation technology called view animation. You should use property animation in new projects.

Overview

The Property Animation API consists of types in the **android.animation** package. The old animation API, called view animation, resides in the **android.view.animation** package. This chapter focuses on the new animation API and does not discuss the older technology. It also does not discuss drawable animation, which is the type of animation that works by loading a series of images, played one after another like a roll of film. For more information on drawable animation, see the documentation for **android.graphics.drawable.AnimationDrawable**.

Property Animation

The powerhouse behind property animation is the **android.animation.Animator** class. It is an abstract class, so you do not use this class directly. Instead, you use one of its subclasses, either **ValueAnimator** or **ObjectAnimator**, to create an animation. In addition, the **AnimatorSet** class, another subclass of **Animator**, is designed to run multiple animations in parallel or sequentially.

All these classes reside in the same package and this section looks at these classes.

Animator

The **Animator** class is an abstract class that provides methods that are inherited by subclasses. There is a method for setting the target object to be animated (**setTarget**), a method for setting the duration (**setDuration**), and a method for starting the animation (**start**). The **start** method can be called more than once on an **Animator** object.

In addition, this class provides an **addListener** method that takes an **Animator.AnimatorListener** instance. The **AnimatorListener** interface is defined inside the **Animator** class and provides methods that will be called by the system upon the occurrence of certain events. You can implement any of these methods if you want to respond to a certain event.

The following are methods in **AnimatorListener**.

```
void onAnimationStart(Animator animation);

void onAnimationEnd(Animator animation);

void onAnimationCancel(Animator animation);

void onAnimationRepeat(Animator animation);
```

For example, the **onAnimationStart** method is called when the animation starts and the **onAnimationEnd** method is called when it ends.

ValueAnimator

A **ValueAnimator** creates an animation by calculating a value that transitions from a start value and to an end value. You specify what the start value and end value should be when constructing the **ValueAnimator**. By registering an **UpdateListener** to a **ValueAnimator**, you can receive an update at each frame, giving you a chance to update your object(s).

Here are two static factory methods that you can use to construct a **ValueAnimator**.

```
public static ValueAnimator ofFloat(float... values)

public static ValueAnimator ofInt(int... values)
```

Which method you should use depends on whether you want to receive an **int** or a **float** in each frame.

Once you create a **ValueAnimator**, you should create an implementation of **AnimationUpdateListener**, write your animation code under its **onAnimationUpdate** method, and register the listener with the **ValueAnimator**. Here is an example.

```
valueAnimator.addUpdateListener(new
        ValueAnimator.AnimatorUpdateListener() {
    @Override
    public void onAnimationUpdate(ValueAnimator animation) {
        Float value = (Float) animation.getAnimatedValue();
        // use value to set a property or multiple properties
        // Example: view.setRotationX(value);
    }
});
```

Finally, call the **ValueAnimator**'s **setDuration** method to set a duration and its **start** method to start the animation. If you do not call **setDuration**, the default method (300ms) will be used.

More on using **ValueAnimator** is given in the example below.

ObjectAnimator

The **ObjectAnimator** class offers the easiest way to animate an object, most probably a **View**, by continually updating one of its properties. To create an animation, create an **ObjectAnimator** using one of its factory methods, passing a target object, a property name, and the start and end values for the property. In recognition of the fact that a property can have an **int** value, a **float** value, or another type of value, **ObjectAnimator** provides three static methods: **ofInt**, **ofFloat**, and **ofObject**. Here are their signatures.

```
public static ObjectAnimator ofInt(java.lang.Object target,
        java.lang.String propertyName, int... values)
```

```
public static ObjectAnimator ofFloat(java.lang.Object target,
        java.lang.String propertyName, float... values)
```

```
public static ObjectAnimator ofObject(java.lang.Object target,
        java.lang.String propertyName, java.lang.Object... values)
```

You can pass one or two arguments to the *values* argument. If you pass two arguments, the first will be used as the start value and the second the end value. If you pass one argument, the value will be used as the end value and the current value of the property will be used as the start value.

Once you have an **ObjectAnimator**, call the **setDuration** method on the **ObjectAnimator** to set the duration and the **start** method to start it. Here is an example of animating the **rotation** property of a **View**.

```
ObjectAnimator objectAnimator = ObjectAnimator.ofFloat(view,
        "rotationY", 0F, 720.0F); // rotate 720 degrees.
objectAnimator.setDuration(2000); // 2000 milliseconds
objectAnimator.start();
```

Running the animation will cause the view to make two full circles within two seconds.

As you can see, you just need two or three lines of code to create a property animation using **ObjectAnimator**. You will learn more about **ObjectAnimator** in the example below.

AnimatorSet

An **AnimatorSet** is useful if you want to play a set of animations in a certain order. A direct subclass of **Animator**, the **AnimatorSet** class allows you to play multiple animations together or one after another. Once you're finished deciding how your animations should be called, call the **start** method on the **AnimatorSet** to start it.

The **playTogether** method arranges the supplied animations to play together. There are two overrides for this method.

```
public void playTogether(java.util.Collection<Animator> items)
```

```
public void playTogether(Animator... items)
```

The **playSequentially** method arranges the supplied animations to play sequentially. It too has two overrides.

```
public void playSequentially(Animator... items)
```

```
public void playSequentially(java.util.List<Animator> items)
```

An Animation Project

The AnimationDemo project uses the **ValueAnimator**, **ObjectAnimator**, and **AnimatorSet** to animate an **ImageView**. It provides three buttons to play different animations.

The manifest for the application is given in Listing 15.1.

Listing 15.1: The manifest for AnimationDemo

```xml
<?xml version="1.0" encoding="utf-8"?>
<manifest xmlns:android="http://schemas.android.com/apk/res/android"
    package="com.example.animationdemo"
    android:versionCode="1"
    android:versionName="1.0" >

    <uses-sdk
        android:minSdkVersion="11"
        android:targetSdkVersion="18" />

    <application
        android:allowBackup="true"
        android:icon="@drawable/ic_launcher"
        android:label="@string/app_name"
        android:theme="@style/AppTheme" >
        <activity
            android:name="com.example.animationdemo.MainActivity"
            android:label="@string/app_name" >
            <intent-filter>
                <action android:name="android.intent.action.MAIN" />

                <category android:name="android.intent.category.LAUNCHER"
    />
            </intent-filter>
        </activity>
    </application>

</manifest>
```

Note that the minimum SDK level is 11 (Honeycomb).

The application has one activity, whose layout is printed in Listing 15.2

Listing 15.2: The activity_main.xml file

```xml
<LinearLayout
    xmlns:android="http://schemas.android.com/apk/res/android"
    xmlns:tools="http://schemas.android.com/tools"
    android:layout_width="match_parent"
    android:layout_height="match_parent"
    android:paddingBottom="@dimen/activity_vertical_margin"
    android:paddingLeft="@dimen/activity_horizontal_margin"
    android:paddingRight="@dimen/activity_horizontal_margin"
    android:paddingTop="@dimen/activity_vertical_margin"
    android:orientation="vertical"
```

```
        tools:context=".MainActivity" >

    <LinearLayout
        android:layout_width="match_parent"
        android:layout_height="wrap_content">

        <Button
            android:id="@+id/button1"
            android:text="@string/button_animate1"
            android:textColor="#ff4433"
            android:layout_width="wrap_content"
            android:layout_height="wrap_content"
            android:onClick="animate1"/>
        <Button
            android:id="@+id/button2"
            android:text="@string/button_animate2"
            android:textColor="#33ff33"
            android:layout_width="wrap_content"
            android:layout_height="wrap_content"
            android:onClick="animate2"/>
        <Button
            android:id="@+id/button3"
            android:text="@string/button_animate3"
            android:textColor="#3398ff"
            android:layout_width="wrap_content"
            android:layout_height="wrap_content"
            android:onClick="animate3"/>

    </LinearLayout>
    <ImageView
        android:id="@+id/imageView1"
        android:layout_width="wrap_content"
        android:layout_height="wrap_content"
        android:layout_gravity="top|center"
        android:src="@drawable/photo1" />
</LinearLayout>
```

The layout defines an **ImageView** and three **Button**s.

Finally, Listing 15.3 shows the **MainActivity** class for the application. There are three event-processing methods (**animate1**, **animate2**, and **animate3**) that each uses a different animation method.

Listing 15.3: The MainActivity class

```
package com.example.animationdemo;
import android.animation.AnimatorSet;
import android.animation.ObjectAnimator;
import android.animation.ValueAnimator;
import android.app.Activity;
import android.os.Bundle;
import android.view.Menu;
import android.view.View;
```

```java
public class MainActivity extends Activity {

    @Override
    protected void onCreate(Bundle savedInstanceState) {
        super.onCreate(savedInstanceState);
        setContentView(R.layout.activity_main);
    }

    @Override
    public boolean onCreateOptionsMenu(Menu menu) {
        getMenuInflater().inflate(R.menu.menu_main, menu);
        return true;
    }

    public void animate1(View source) {
        View view = findViewById(R.id.imageView1);
        ObjectAnimator objectAnimator = ObjectAnimator.ofFloat(
                view, "rotationY", 0F, 720.0F);
        objectAnimator.setDuration(2000);
        objectAnimator.start();
    }

    public void animate2(View source) {
        final View view = findViewById(R.id.imageView1);
        ValueAnimator valueAnimator = ValueAnimator.ofFloat(0F,
                7200F);
        valueAnimator.setDuration(15000);

        valueAnimator.addUpdateListener(new
                ValueAnimator.AnimatorUpdateListener() {
            @Override
            public void onAnimationUpdate(ValueAnimator animation) {
                Float value = (Float) animation.getAnimatedValue();
                view.setRotationX(value);
                if (value < 3600) {
                    view.setTranslationX(value/20);
                    view.setTranslationY(value/20);
                } else {
                    view.setTranslationX((7200-value)/20);
                    view.setTranslationY((7200-value)/20);
                }
            }
        });
        valueAnimator.start();
    }
    public void animate3(View source) {
        View view = findViewById(R.id.imageView1);
        ObjectAnimator objectAnimator1 =
                ObjectAnimator.ofFloat(view, "translationY", 0F,
                        300.0F);
        ObjectAnimator objectAnimator2 =
                ObjectAnimator.ofFloat(view, "translationX", 0F,
                        300.0F);
```

```
        objectAnimator1.setDuration(2000);
        objectAnimator2.setDuration(2000);
        AnimatorSet animatorSet = new AnimatorSet();
        animatorSet.playTogether(objectAnimator1, objectAnimator2);

        ObjectAnimator objectAnimator3 =
                ObjectAnimator.ofFloat(view, "rotation", 0F,
                        1440F);
        objectAnimator3.setDuration(4000);
        animatorSet.play(objectAnimator3).after(objectAnimator2);
        animatorSet.start();
    }
}
```

Run the application and click the buttons to play the animations. Figure 15.1 shows the application.

Figure 15.1: Animation demo

Summary

In this chapter you learned about the new Animation API in Android, the Property Animation system. In particular, you learned about the **android.animation.Animator** class and its subclasses, **ValueAnimator** and **ObjectAnimator**. You also learned to use the **AnimatorSet** class to perform multiple animations.

Chapter 16
Preferences

Android comes with a **SharedPreferences** interface that can be used to manage application settings as key/value pairs. **SharedPreferences** also takes care of the writing of data to a file. In addition, Android provides the Preference API with user interface (UI) classes that are linked to the default **SharedPreferences** instance so that you can easily create a UI for modifying application settings.

This chapter discusses **SharedPreferences** and the Preference API in detail.

SharedPreferences

The **android.content.SharedPreferences** interface provides methods for storing and reading application settings. You can obtain the default instance of **SharedPreferences** by calling the **getDefaultSharedPreferences** static method of **PreferenceManager**, passing a **Context**.

```
PreferenceManager.getDefaultSharedPreferences(context);
```

To read a value from the **SharedPreferences**, use one of the following methods.

```
public int getInt(java.lang.String key, int default)
```

```
public boolean getBoolean(java.lang.String key, boolean default)
```

```
public float getFloat(java.lang.String key, float default)
```

```
public long getLong(java.lang.String key, long default)
```

```
public int getString(java.lang.String key, java.lang.String default)
```

```
public java.util.Set<java.lang.String> getStringSet(
    java.lang.String key, java.util.Set<java.lang.String> default)
```

The **getXXX** methods return the value associated with the specified key if the pair exists. Otherwise, it returns the specified default value.

To first check if a **SharedPreferences** contains a key/value pair, use the **contains** method, which returns **true** if the specified key exists.

```
public boolean contains(java.lang.String key)
```

On top of that, you can use the **getAll** method to get all key-value pairs as a **Map**.

```
public java.util.Map<java.lang.String, ?> getAll()
```

Values stored in a **SharedPreferences** are persisted automatically and will survive user sessions. The values will be deleted when the application is uninstalled.

The Preference API

To store a key-value pair in a **SharedPreferences**, you normally use the Android Preference API to create a user interface that the user can use to edit settings. The **android.preference.Preference** class is the main class for this. Some of its subclasses are

- **CheckBoxPreference**
- **EditTextPreference**
- **ListPreference**
- **DialogPreference**

An instance of a **Preference** subclass corresponds to a setting.

You could create a **Preference** at runtime, but the best way to create one is by using an XML file to lay out your preferences and then use a **PreferenceFragment** to load the XML file. The XML file must have a **PreferenceScreen** root element and is commonly named **preferences.xml** and should be saved to an **xml** directory under **res**.

Note
Prior to Android 3.0, the **PreferenceActivity** was often used to load a preference xml file. This class is now deprecated and should not be used. Use **PreferenceFragment**, instead.

You will learn how to use **Preference** in the following example.

Using Preferences

The PreferenceDemo1 application shows you how you can use **SharedPreferences** and the Preference API. It has two activities. The first activity shows the values of three application settings by reading them when the activity is resumed. The second activity contains a **PreferenceFragment** that allows the user to change each of the settings.

Figures 16.1 and 16.2 show the main activity and the second activity, respectively.

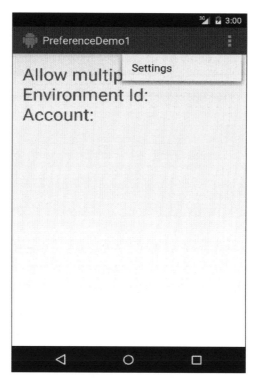

Figure 16.1: The main activity of PreferenceDemo1

Figure 16.2: The SettingsActivity activity

The **AndroidManifest.xml** file for the application, which describes the two activities, is shown in Listing 16.1.

Listing 16.1: The AndroidManifest.xml file

```xml
<?xml version="1.0" encoding="utf-8"?>
<manifest xmlns:android="http://schemas.android.com/apk/res/android"
    package="com.example.preferencedemo1"
    android:versionCode="1"
    android:versionName="1.0" >

    <uses-sdk
        android:minSdkVersion="19"
        android:targetSdkVersion="19" />

    <application
        android:allowBackup="true"
        android:icon="@drawable/ic_launcher"
        android:label="@string/app_name"
        android:theme="@style/AppTheme" >
        <activity
            android:name="com.example.preferencedemo1.MainActivity"
            android:label="@string/app_name">
            <intent-filter>
```

```
                <action android:name="android.intent.action.MAIN"/>
                <category
android:name="android.intent.category.LAUNCHER" />
            </intent-filter>
        </activity>
        <activity
android:name="com.example.preferencedemo1.SettingsActivity"
            android:parentActivityName=".MainActivity"
            android:label="">
        </activity>
    </application>
</manifest>
```

The first activity has a very simple layout that features a sole **TextView** as is presented by the **activity_main.xml** file in Listing 16.2.

Listing 16.2: The layout file for the first activity (activity_main.xml)

```
<RelativeLayout
    xmlns:android="http://schemas.android.com/apk/res/android"
    android:layout_width="match_parent"
    android:layout_height="match_parent"
    android:paddingBottom="@dimen/activity_vertical_margin"
    android:paddingLeft="@dimen/activity_horizontal_margin"
    android:paddingRight="@dimen/activity_horizontal_margin"
    android:paddingTop="@dimen/activity_vertical_margin">
    <TextView
        android:id="@+id/info"
        android:layout_width="wrap_content"
        android:layout_height="wrap_content"
        android:textSize="30sp"/>
</RelativeLayout>
```

The **MainActivity** class, shown in Listing 16.3, is the activity class for the first activity. It reads three settings from the default **SharedPreferences** in its **onResume** method and displays the values in the **TextView**.

Listing 16.3: The MainActivity class

```
package com.example.preferencedemo1;
import android.app.Activity;
import android.content.Intent;
import android.content.SharedPreferences;
import android.os.Bundle;
import android.preference.PreferenceManager;
import android.view.Menu;
import android.view.MenuItem;
import android.widget.TextView;

public class MainActivity extends Activity {

    @Override
    protected void onCreate(Bundle savedInstanceState) {
        super.onCreate(savedInstanceState);
        setContentView(R.layout.activity_main);
```

```
    }

    @Override
    public void onResume() {
        super.onResume();
        SharedPreferences sharedPref = PreferenceManager.
                getDefaultSharedPreferences(this);
        boolean allowMultipleUsers = sharedPref.getBoolean(
                SettingsActivity.ALLOW_MULTIPLE_USERS, false);
        String envId = sharedPref.getString(
                SettingsActivity.ENVIRONMENT_ID, "");
        String account = sharedPref.getString(
                SettingsActivity.ACCOUNT, "");
        TextView textView = (TextView) findViewById(R.id.info);
        textView.setText("Allow multiple users: " +
                allowMultipleUsers + "\nEnvironment Id: " + envId
                + "\nAccount: " + account);
    }

    @Override
    public boolean onCreateOptionsMenu(Menu menu) {
        getMenuInflater().inflate(R.menu.menu_main, menu);
        return true;
    }

    @Override
    public boolean onOptionsItemSelected(MenuItem item) {
        switch (item.getItemId()) {
            case R.id.action_settings:
                startActivity(new Intent(this,
                        SettingsActivity.class));
                return true;
            default:
                return super.onOptionsItemSelected(item);
        }
    }
}
```

In addition, the **MainActivity** class overrides the **onCreateOptionsMenu** and **onOptionsItemSelected** methods so that a Settings action appears on the action bar and clicking it will start the second activity, **SettingsActivity**.

SettingsActivity, presented in Listing 16.4, contains a default layout that is replaced by an instance of **SettingsFragment** when the activity is created. Pay attention to the **onCreate** method of the class. If the last lines of code in the method looks foreign to you, please first read Chapter 13, "Fragments."

Listing 16.4: The SettingsActivity class

```
package com.example.preferencedemo1;
import android.app.Activity;
import android.os.Bundle;
import android.view.Menu;
```

```
public class SettingsActivity extends Activity {

    public static final String ALLOW_MULTIPLE_USERS =
            "allowMultipleUsers";
    public static final String ENVIRONMENT_ID = "envId";
    public static final String ACCOUNT = "account";

    @Override
    protected void onCreate(Bundle savedInstanceState) {
        super.onCreate(savedInstanceState);
        getActionBar().setDisplayHomeAsUpEnabled(true);
        getFragmentManager()
                .beginTransaction()
                .replace(android.R.id.content,
                        new SettingsFragment()).commit();
    }

    @Override
    public boolean onCreateOptionsMenu(Menu menu) {
        getMenuInflater().inflate(R.menu.menu_settings, menu);
        return true;
    }
}
```

Note that the **SettingsActivity** class declares three public static finals that define setting keys. The fields are used internally as well as from other classes.

The **SettingsFragment** class is a subclass of **PreferenceFragment**. It is a simple class that simply calls **addPreferencesFromResource** to load the XML document containing the layout for three Preference subclasses. The **SettingsFragment** class is shown in Listing 16.5 and the XML file in Listing 16.6.

Listing 16.5: The SettingsFragment class

```
package com.example.preferencedemo1;
import android.os.Bundle;
import android.preference.PreferenceFragment;

public class SettingsFragment extends PreferenceFragment {

    @Override
    public void onCreate(Bundle savedInstanceState) {
        super.onCreate(savedInstanceState);

        // Load the preferences from an XML resource
        addPreferencesFromResource(R.xml.preferences);
    }
}
```

Listing 16.6: The res/xml/preferences.xml file

```
<PreferenceScreen
        xmlns:android="http://schemas.android.com/apk/res/android">

    <PreferenceCategory android:title="Category 1">
        <CheckBoxPreference
```

```
            android:key="allowMultipleUsers"
            android:title="Allow multiple users"
            android:summary="Allow multiple users" />
    </PreferenceCategory>

    <PreferenceCategory android:title="Category 2">
        <EditTextPreference
            android:key="envId"
            android:title="Environment Id"
            android:dialogTitle="Environment Id"/>
        <EditTextPreference
            android:key="account"
            android:title="Account"/>
    </PreferenceCategory>
</PreferenceScreen>
```

The **preferences.xml** file groups the **Preference** subclasses into two categories. In the first category is a **CheckBoxPreference** linked to the **allowMultipleUsers** key. In the second category are two **EditTextPreference**s linked to **envId** and **account**.

Summary

An easy way to manage application settings is by using the Preference API and the default **SharedPreferences**. In this chapter you learned how to use both.

Chapter 17
Working with Files

Reading from and writing to a file are some of the most common operations in any type of application, including Android. In this chapter, you will learn how Android structures its storage areas and how to use the Android File API.

Overview

Android devices offer two storage areas, internal and external. The internal storage is private to the application. The user and other applications cannot access it.

The external storage is where you store files that will be shared with other applications or that the user will be able to access. For example, the built-in Camera application stores digital image files in the external storage so the user can easily copy them to a computer.

Internal Storage

All applications can read from and write to internal storage. The location of the internal storage is **/data/data/[*app package*]**, so if your application package is **com.example.myapp**, the internal directory for this application is **/data/data/com.example.myapp**. The **Context** class provides various methods to access the internal storage from your application. You should use these methods to access files you store in the internal storage and should not hardcode the location of the internal storage. (Recall that **Activity** is a subclass of **Context**, so you can call public and protected methods in **Context** from your activity class). Here are the methods in **Context** for working with files and streams in the internal storage.

```
public java.io.File getFilesDir()
```
Returns the path to the directory dedicated to your application in internal storage.

```
public java.io.FileOutputStream openFileOutput(
        java.lang.String name, int mode)
```
Opens a **FileOutputStream** in the application's section of internal storage.

```
public java.io.FileInputStream openFileInput(java.lang.String name)
```
Opens a **FileInputStream** for reading. The name argument is the name of the file to open and cannot contain path separators.

```
public java.io.File getFilesDir()
```
Obtains the absolute path to the file system directory where internal files are saved.

```
public java.io.File getDir(java.lang.String name, int mode)
```
Creates or retrieves an existing directory within the application's internal storage space. The **name** argument is the name of the directory to retrieve and the **mode** argument should be given one of these:
MODE_PRIVATE for the default operation or **MODE_WORLD_READABLE** or **MODE_WORLD_WRITEABLE** to control permissions.

```
public boolean deleteFile(java.lang.String fileName)
```
Deletes a file saved on the internal storage. The method returns **true** if the file was successfully deleted.

```
public java.lang.String[] FileList()
```
Returns an array of strings naming the files associated with this Context's application package.

External Storage

There are two types of files that can be written to external storage, private files and public files. Private files are private to the application and will be deleted when the application is uninstalled. Public files, on the other hand, are meant to be shared with other applications or accessible to the user.

External storage may be removable. As such, there is a difference between files stored in internal storage and files stored in external storage as public files. Files in internal storage are secure and cannot be accessed by the user or other applications. Public files in external storage do not enjoy the same level of security as the user can remove the storage and use some tool to access the files.

Since external storage can be removed, when you try to read from or write to it, you should first test if external storage is available. Trying to access external storage when it is unavailable may crash your application.

To inquire if external storage is available, use one of these methods.

```
public boolean isExternalStorageWritable() {
    String state = Environment.getExternalStorageState();
    return Environment.MEDIA_MOUNTED.equals(state);
}

public boolean isExternalStorageReadable() {
    String state = Environment.getExternalStorageState();
    return (Environment.MEDIA_MOUNTED.equals(state) ||
            Environment.MEDIA_MOUNTED_READ_ONLY.equals(state));
}
```

You can use the **getExternalFilesDir** method on the **Context** to get the directory for storing private files on the external storage.

Public files can be stored in the directory returned by the **getExternalStoragePublicDirectory** method of the **android.os.Environment** class. Here is the signature of this method.

```
public static java.io.File getExternalStoragePublicDirectory(
        java.lang.String type)
```

Here, *type* is a directory under the root directory. The **Environment** class provides the following fields that you can use for various file types.

- **Directory_ALARMS**
- **Directory_DCIM**
- **Directory_DOCUMENTS**
- **Directory_DOWNLOADS**
- **Directory_MOVIES**
- **Directory_MUSIC**
- **Directory_NOTIFICATIONS**
- **Directory_PICTURES**
- **Directory_PODCASTS**
- **Directory_RINGTONES**

For example, music files should be stored in the directory returned by this code.

```
File dir = Environment.getExternalStoragePublicDirectory(
        Environment.DIRECTORY_PICTURES)
```

Writing to external storage requires user permission. To ask the user to grant you read and write access to external storage, add this in your manifest.

```
<uses-permission
        android:name="android.permission.WRITE_EXTERNAL_STORAGE"/>
```

Currently, if your application only needs to read from external storage, you do not need special permissions. However, this will change in the future so you should declare this **uses-permission** element in your manifest if you need to read from external storage so that your application will keep working after the change takes effect.

```
<uses-permission
        android:name="android.permission.READ_EXTERNAL_STORAGE"/>
```

Creating a Notes Application

The FileDemo1 application is a simple application for managing notes. A note has a title and a body and each note is stored as a file, using the title as the file name. The user can view a list of notes, view a note, create a new note, and delete a note.

The application has two activities, **MainActivity** and **AddNoteActivity**. The **MainActivity** activity uses a **ListView** that lists all note titles in the system. The main activity contains a **ListView** that lists all note titles. Selecting a note title from the **ListView** shows the note in the **TextView** beside the **ListView**.

Figures 17.1 and 17.2 show the **MainActivity** activity and **AddNoteActivity** activity, respectively. The **MainActivity** activity contains a **ListView** that lists all note titles and a **TextView** that shows the body of the selected note. Its action bar also contains two buttons, Add and Delete. Add starts the **AddNoteActivity** activity. Delete deletes the selected note.

Figure 17.1: MainActivity

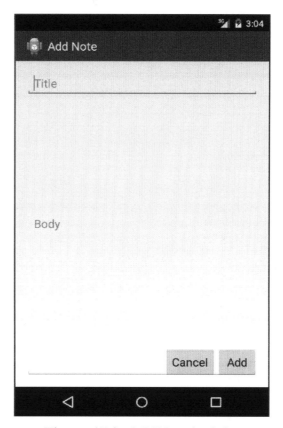

Figure 17.2: AddNoteActivity

Let's now take a look at the application code. Listing 17.1 shows the
AndroidManifest.xml file for this application.

Listing 17.1: The AndroidManifest.xml file

```
<?xml version="1.0" encoding="utf-8"?>
<manifest xmlns:android="http://schemas.android.com/apk/res/android"
    package="com.example.filedemo1"
    android:versionCode="1"
    android:versionName="1.0" >

    <uses-sdk
        android:minSdkVersion="19"
        android:targetSdkVersion="19" />

    <application
        android:allowBackup="true"
        android:icon="@drawable/ic launcher"
        android:label="@string/app name"
        android:theme="@style/AppTheme" >
        <activity
            android:name="com.example.filedemo1.MainActivity"
            android:label="@string/app_name" >
```

```
        <intent-filter>
            <action android:name="android.intent.action.MAIN"/>
            <category
android:name="android.intent.category.LAUNCHER"/>
        </intent-filter>
    </activity>
    <activity
        android:name="com.example.filedemo1.AddNoteActivity"
        android:label="@string/title_activity_add_note" >
    </activity>
  </application>
</manifest>
```

The manifest declares the two activities in the application. The activity class of the main activity, **MainActivity**, is shown in Listing 17.2.

Listing 17.2: The MainActivity class

```java
package com.example.filedemo1;
import java.io.BufferedReader;
import java.io.File;
import java.io.FileReader;
import java.io.IOException;
import android.app.Activity;
import android.content.Intent;
import android.os.Bundle;
import android.view.Menu;
import android.view.MenuItem;
import android.view.View;
import android.widget.AdapterView;
import android.widget.AdapterView.OnItemClickListener;
import android.widget.ArrayAdapter;
import android.widget.ListView;
import android.widget.TextView;

public class MainActivity extends Activity {
    private String selectedItem;

    @Override
    protected void onCreate(Bundle savedInstanceState) {
        super.onCreate(savedInstanceState);
        setContentView(R.layout.activity_main);
        ListView listView = (ListView) findViewById(
                R.id.listView1);
        listView.setChoiceMode(ListView.CHOICE_MODE_SINGLE);
        listView.setOnItemClickListener(
                new OnItemClickListener() {
            @Override
            public void onItemClick(AdapterView<?> adapterView,
                    View view, int position, long id) {
                readNote(position);
            }
        });
    }
```

```java
@Override
public void onResume() {
    super.onResume();
    refreshList();
}

@Override
public boolean onCreateOptionsMenu(Menu menu) {
    getMenuInflater().inflate(R.menu.menu_main, menu);
    return true;
}
@Override
public boolean onOptionsItemSelected(MenuItem item) {
    // Handle presses on the action bar items
    switch (item.getItemId()) {
        case R.id.action_add:
            startActivity(new Intent(this,
                    AddNoteActivity.class));
            return true;
        case R.id.action_delete:
            deleteNote();
            return true;
        default:
            return super.onOptionsItemSelected(item);
    }
}

private void refreshList() {
    ListView listView = (ListView) findViewById(
            R.id.listView1);
    String[] titles = fileList();
    ArrayAdapter<String> arrayAdapter =
            new ArrayAdapter<String>(
            this,
            android.R.layout.simple_list_item_activated_1,
            titles);
    listView.setAdapter(arrayAdapter);
}

private void readNote(int position) {
    String[] titles = fileList();
    if (titles.length > position) {
        selectedItem = titles[position];
        File dir = getFilesDir();
        File file = new File(dir, selectedItem);
        FileReader fileReader = null;
        BufferedReader bufferedReader = null;
        try {
            fileReader = new FileReader(file);
            bufferedReader = new BufferedReader(fileReader);
            StringBuilder sb = new StringBuilder();
            String line = bufferedReader.readLine();
```

```
                while (line != null) {
                    sb.append(line);
                    line = bufferedReader.readLine();
                }
                ((TextView) findViewById(R.id.textView1)).
                        setText(sb.toString());
            } catch (IOException e) {

            } finally {
                if (bufferedReader != null) {
                    try {
                        bufferedReader.close();
                    } catch (IOException e) {
                    }
                }
                if (fileReader != null) {
                    try {
                        fileReader.close();
                    } catch (IOException e) {
                    }
                }
            }
        }
    }

    private void deleteNote() {
        if (selectedItem != null) {
            deleteFile(selectedItem);
            selectedItem = null;
            ((TextView) findViewById(R.id.textView1)).setText("");
            refreshList();
        }
    }
}
```

The **MainActivity** class contains a **ListView** and its **onCreate** method sets the **ListView**'s choice mode and passes a listener to it.

```
ListView listView = (ListView) findViewById(
        R.id.listView1);
listView.setChoiceMode(ListView.CHOICE_MODE_SINGLE);
listView.setOnItemClickListener( ... )
```

No list adapter is passed to the **ListView** in the **onCreate** method. Instead, the **onResume** method calls the **refreshNotes** method, which passes a new **ListAdapter** to the **ListView** every time **onResume** is called. The reason why a new **ListAdapter** needs to be created every time **onResume** is called is because the main activity can call the **AddNoteActivity** for the user to add a note. If the user does add a note and leave the **AddNoteActivity**, the main activity needs to include the new note, hence the need to refresh the **ListView**.

Note
Automatic refresh in a **ListView** can be done using a Cursor. See Chapter 18, "Working with the Database."

The **readNote** method, which gets called when a list item is selected, starts by getting all file names in the internal storage.

```
String[] titles = fileList();
```

It then retrieves the note title and uses it to create a file, using the directory returned by **getFilesDir** as the parent.

```
if (titles.length > position) {
    selectedItem = titles[position];
    File dir = getFilesDir();
    File file = new File(dir, selectedItem);
```

The **readNote** method then uses a **FileReader** and a **BufferedReader** to read the note, one line at a time, and sets the value of the **TextView**.

```
FileReader fileReader = null;
BufferedReader bufferedReader = null;
try {
    fileReader = new FileReader(file);
    bufferedReader = new BufferedReader(fileReader);
    StringBuilder sb = new StringBuilder();
    String line = bufferedReader.readLine();
    while (line != null) {
        sb.append(line);
        line = bufferedReader.readLine();
    }
    ((TextView) findViewById(R.id.textView1)).
            setText(sb.toString());
```

The **AddNoteActivity** class is shown in Listing 17.3.

Listing 17.3: The AddNoteActivity class

```
package com.example.filedemo1;
import java.io.File;
import java.io.PrintWriter;
import android.app.Activity;
import android.app.AlertDialog;
import android.os.Bundle;
import android.view.View;
import android.widget.EditText;

public class AddNoteActivity extends Activity {

    @Override
    protected void onCreate(Bundle savedInstanceState) {
        super.onCreate(savedInstanceState);
        setContentView(R.layout.activity_add_note);
    }

    public void cancel(View view) {
        finish();
    }
```

```
public void addNote(View view) {
    String fileName = ((EditText)
            findViewById(R.id.noteTitle))
        .getText().toString();
    String body = ((EditText) findViewById(R.id.noteBody))
        .getText().toString();
    File parent = getFilesDir();
    File file = new File(parent, fileName);
    PrintWriter writer = null;
    try {
        writer = new PrintWriter(file);
        writer.write(body);
        finish();
    } catch (Exception e) {
        showAlertDialog("Error adding note", e.getMessage());
    } finally {
        if (writer != null) {
            try {
                writer.close();
            } catch (Exception e) {

            }
        }
    }
}

private void showAlertDialog(String title, String message) {
    AlertDialog alertDialog = new
            AlertDialog.Builder(this).create();
    alertDialog.setTitle(title);
    alertDialog.setMessage(message);
    alertDialog.show();
}
}
```

The **AddNoteActivity** class has two public methods that act as click listeners to the two buttons in its layout, cancel and addNote. The **cancel** method simply closes the activity. The **addNote** method reads the values in the **TextView**s and create a file in the internal storage using a **PrintWriter**.

Accessing the Public Storage

The second example in this chapter, FileDemo2, shows how you can access the public storage. FileDemo2 is a file browser that shows the content of a standard directory. There is only one activity in FileDemo2 and it is shown in Figure 17.3.

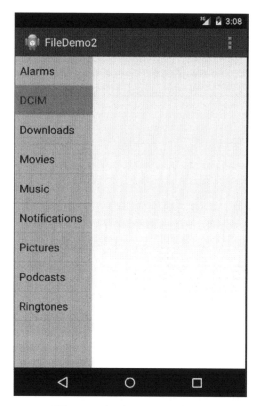

Figure 17.3: FileDemo2

The layout file for the activity is presented in Listing 17.4. It is a **LinearLayout** that contains two **ListView**s. The **ListView** on the left lists several frequently used directories. The **ListView** on the right shows the content of the selected directory.

Listing 17.4: The layout file for FileDemo2's activity

```
<LinearLayout xmlns:android="http://schemas.android.com/apk/res/android"
    android:layout_width="match_parent"
    android:layout_height="match_parent"
    android:orientation="horizontal">
    <ListView
        android:id="@+id/listView1"
        android:layout_width="0sp"
        android:layout_weight="1"
        android:layout_height="match_parent"
        android:background="#ababff"/>
    <ListView
        android:id="@+id/listView2"
        android:layout_width="0sp"
        android:layout_height="wrap_content"
        android:layout_weight="2"/>
</LinearLayout>
```

The activity class for FileDemo2 is shown in Listing 17.5.

Listing 17.5: The MainActivity class

```java
package com.example.filedemo2;
import java.io.File;
import java.util.Arrays;
import java.util.List;

import android.app.Activity;
import android.os.Bundle;
import android.os.Environment;
import android.view.Menu;
import android.view.View;
import android.widget.AdapterView;
import android.widget.AdapterView.OnItemClickListener;
import android.widget.ArrayAdapter;
import android.widget.ListView;

public class MainActivity extends Activity {
    class KeyValue {
        public String key;
        public String value;
        public KeyValue(String key, String value) {
            this.key = key;
            this.value = value;
        }
        @Override
        public String toString() {
            return key;
        }
    }

    @Override
    protected void onCreate(Bundle savedInstanceState) {
        super.onCreate(savedInstanceState);
        setContentView(R.layout.activity_main);
        final List<KeyValue> keyValues = Arrays.asList(
            new KeyValue("Alarms", Environment.DIRECTORY_ALARMS),
            new KeyValue("DCIM", Environment.DIRECTORY_DCIM),
            new KeyValue("Downloads",
                    Environment.DIRECTORY_DOWNLOADS),
            new KeyValue("Movies", Environment.DIRECTORY_MOVIES),
            new KeyValue("Music", Environment.DIRECTORY_MUSIC),
            new KeyValue("Notifications",
                    Environment.DIRECTORY_NOTIFICATIONS),
            new KeyValue("Pictures",
                    Environment.DIRECTORY_PICTURES),
            new KeyValue("Podcasts",
                    Environment.DIRECTORY_PODCASTS),
            new KeyValue("Ringtones",
                    Environment.DIRECTORY_RINGTONES)
        );
        ArrayAdapter<KeyValue> arrayAdapter = new
                ArrayAdapter<KeyValue>(this,
```

```
                    android.R.layout.simple_list_item_activated_1,
                    keyValues);
        ListView listView1 = (ListView)
                findViewById(R.id.listView1);
        listView1.setChoiceMode(ListView.CHOICE_MODE_SINGLE);
        listView1.setAdapter(arrayAdapter);
        listView1.setOnItemClickListener(new
                OnItemClickListener() {
            @Override
            public void onItemClick(AdapterView<?> adapterView,
                    View view, int position, long id) {
                KeyValue keyValue = keyValues.get(position);
                listDir(keyValue.value);
            }
        });
    }


    @Override
    public boolean onCreateOptionsMenu(Menu menu) {
        getMenuInflater().inflate(R.menu.menu_main, menu);
        return true;
    }

    private void listDir(String dir) {
        File parent = Environment
                .getExternalStoragePublicDirectory(dir);
        String[] files = null;
        if (parent == null || parent.list() == null) {
            files = new String[0];
        } else {
            files = parent.list();
        }
        ArrayAdapter<String> arrayAdapter = new
                ArrayAdapter<String>(this,
                    android.R.layout.simple_list_item_activated_1,
                    files);
        ListView listView2 = (ListView)
                findViewById(R.id.listView2);
        listView2.setAdapter(arrayAdapter);
    }
}
```

The first thing to note is the **KeyValue** class in the activity class. This is a simple class to
hold a pair of strings. It is used in the **onCreate** method to pair selected keys with
directories defined in the **Environment** class.

```
            new KeyValue("Alarms", Environment.DIRECTORY_ALARMS),
            new KeyValue("DCIM", Environment.DIRECTORY_DCIM),
            new KeyValue("Downloads",
                    Environment.DIRECTORY_DOWNLOADS),
            new KeyValue("Movies", Environment.DIRECTORY_MOVIES),
            new KeyValue("Music", Environment.DIRECTORY_MUSIC),
            new KeyValue("Notifications",
```

```
                    Environment.DIRECTORY_NOTIFICATIONS),
            new KeyValue("Pictures",
                    Environment.DIRECTORY_PICTURES),
            new KeyValue("Podcasts",
                    Environment.DIRECTORY_PODCASTS),
            new KeyValue("Ringtones",
                    Environment.DIRECTORY_RINGTONES)
```

These **KeyValue** instances are then used to feed the first **ListView**.

```
ArrayAdapter<KeyValue> arrayAdapter = new
        ArrayAdapter<KeyValue>(this,
                android.R.layout.simple_list_item_activated_1,
                keyValues);
ListView listView1 = (ListView)
        findViewById(R.id.listView1);
listView1.setChoiceMode(ListView.CHOICE_MODE_SINGLE);
listView1.setAdapter(arrayAdapter);
```

The **ListView** also gets a listener that listens for its **OnItemClick** event and calls the **listDir** method when one of the directories in the **ListView** is selected.

```
listView1.setOnItemClickListener(new
        OnItemClickListener() {
    @Override
    public void onItemClick(AdapterView<?> adapterView,
            View view, int position, long id) {
        KeyValue keyValue = keyValues.get(position);
        listDir(keyValue.value);
    }
});
```

The **listDir** method list all files in the selected directory and feed them to an **ArrayAdapter** that in turn gets passed to the second **ListView**.

```
private void listDir(String dir) {
    File parent = Environment
            .getExternalStoragePublicDirectory(dir);
    String[] files = null;
    if (parent == null || parent.list() == null) {
        files = new String[0];
    } else {
        files = parent.list();
    }
    ArrayAdapter<String> arrayAdapter = new
            ArrayAdapter<String>(this,
                    android.R.layout.simple_list_item_activated_1,
                    files);
    ListView listView2 = (ListView)
            findViewById(R.id.listView2);
    listView2.setAdapter(arrayAdapter);
}
```

Summary

You use the File API to work with files in Android applications. In addition to mastering this API, in order to work with files effectively in Android, you need to know how Android structures its storage system and the file-related methods defined in the **Context** and **Environment** classes.

Chapter 18
Working with the Database

Android has its own technology for working with databases and it has nothing to do with Java Database Connectivity (JDBC), the technology Java developers use for accessing data in a relational database. In addition, Android ships with SQLite, an open source database.

This chapter shows how to work with the Android Database API and the SQLite database.

Overview

Android comes with its own Database API. The API consists of two packages, **android.database** and **android.database.sqlite**. Android ships with SQLite, an open source relational database that partially implements SQL-92, the third revision of the SQL standard.

Currently at version 3, SQLite offers a minimum number of data types: Integer, Real, Text, Blob, and Numeric. One interesting feature of SQLite is that an integer primary key is automatically auto-incremented when a row is inserted without passing a value for the field.

More information on SQLite can be found here:

```
http://sqlite.org
```

The Database API

The **SQLiteDatabase** and **SQLiteOpenHelper** classes, both part of **android.database.sqlite**, are the two most frequently used classes in the Database API. In the **android.database** package, the **Cursor** interface is one of the most important types.

The three types are discussed in detail in the following subsections.

The SQLiteOpenHelper Class

To use a database in your Android application, extend **SQLiteOpenHelper** to help with database and table creation as well as connecting to the database. In a subclass of **SQLiteOpenHelper**, you need to do the following.

- Provide a constructor that calls its super, passing, among others, the **Context** and the database name.
- Override the **onCreate** and **onUpgrade** methods.

For example, here is a constructor for a subclass of **SQLiteOpenHelper**.

```
public SubClassOfSQLiteOpenHelper(Context context) {
    super(context,
        "mydatabase", // database name
        null,
        1              // db version
    );
}
```

The **onCreate** method that needs to be overridden has the following signature.

```
public void onCreate(SQLiteDatabase database)
```

The system will call **onCreate** the first time access to one of the tables is required. In this method implementation you should call the **execSQL** method on the **SQLiteDatabase** and pass an SQL statement for creating your table(s). Here is an example.

```
@Override
public void onCreate(SQLiteDatabase db) {
    String sql = "CREATE TABLE " + TABLE_NAME
        + " (" + ID_FIELD + " INTEGER, "
        + FIRST_NAME_FIELD + " TEXT,"
        + LAST_NAME_FIELD + " TEXT,"
        + PHONE_FIELD + " TEXT,"
        + EMAIL_FIELD + " TEXT,"
        + " PRIMARY KEY (" + ID_FIELD + "));";
    db.execSQL(sql);
}
```

SQLiteOpenHelper automatically manages connections to the underlying database. To retrieve the database instance, call one of these methods, both of which return an instance of **SQLiteDatabase**.

```
public SQLiteDatabase getReadableDatabase()
```

```
public SQLiteDatabase getWritableDatabase()
```

The first time one of these methods is called a database will be created if none exists. The difference between **getReadableDatabase** and **getWritableDatabase** is the former can be used for read-only whereas the latter can be used to read from and write to the database.

The SQLiteDatabase Class

Once you get a **SQLiteDatabase** from a **SQLiteOpenHelper**'s **getReadableDatabase** or **getWritableDatabase** method, you can manipulate the data in the database by calling the **SQLiteDatabase**'s **insert** or **execSQL** method. For example, to add a record, call the **insert** method whose signature is as follows.

```
public long insert (String table, String nullColumnHack,
        ContentValues values)
```

Here, *table* is the name of the table and *values* is an **android.content.ContentValues** that contains pairs of field names/values to be inserted to the table. This method returns the row identifier for the new row.

For instance, the following code inserts a record into the **employees** table passing three field values.

```
SQLiteDatabase db = this.getWritableDatabase();
// this is an instance of SQLiteOpenHelper
ContentValues values = new ContentValues();
values.put("first_name", "Joe");
values.put("last_name", "Average");
values.put("position", "System Analyst");
long id = db.insert("employees", null, values);
db.close();
```

To update or delete a record, use the **update** or **delete** method, respectively. The signatures of these methods are as follows.

```
public int delete (java.lang.String table,
        java.lang.String whereClause, java.lang.String[] whereArgs)
```

```
public int update (java.lang.String table,
        android.content.ContentValues values,
        java.lang.String whereClause, java.lang.String[] whereArgs)
```

Examples of the two methods are shown in the accompanying application.

To execute a SQL statement, use the **execSQL** method.

```
public void execSQL (java.lang.String sql)
```

Finally, to retrieve records, use one of the **query** methods. One of the method overloads has this signature.

```
public android.database.Cursor query(java.lang.String table,
        java.lang.String[] columns, java.lang.String selection,
        java.lang.String[] selectionArgs,
        java.lang.String groupBy,
        java.lang.String having,
        java.lang.String orderBy, hava.lang.String limit)
```

You can find an example on how to use this method in the sample application accompanying this chapter.

One thing to note: The data returned by the query method is contained in an instance of **Cursor**, an interesting type explained in the next section.

The Cursor Interface

Calling the **query** method on a **SQLiteDatabase** returns a **Cursor**. A **Cursor**, an implementation of the **android.database.Cursor** interface, provides read and write access to the result set returned by a database query.

To read a row of data through a **Cursor**, you first need to point the **Cursor** to a data row by calling its **moveToFirst**, **moveToNext**, **moveToPrevious**, **moveToLast**, or **moveToPosition** method. **moveToFirst** moves the **Cursor** to the first row and **moveToNext** to the next row. **moveToLast**, you may have guessed correctly, moves it to the last record and **moveToPrevious** to the previous row. **moveToPosition** takes an integer and moves the **Cursor** to the specified position.

Once you move the **Cursor** to a data row, you can read a column value in the row by calling the **Cursor**'s **getInt**, **getFloat**, **getLong**, **getString**, **getShort**, or **getDouble** method, passing the column index.

An interesting aspect of **Cursor** is that it can be used as the data source for a **ListAdapter**, which in turn can be used to feed a **ListView**. The advantage of using a **Cursor** for a **ListView** is that the **Cursor** can manage your data. In other words, if the data is updated, the **Cursor** can self-refresh the **ListView**. This is a very useful feature as you then have one fewer thing to worry about.

Example

The DatabaseDemo1 application is an application for managing contacts in a SQLite database. A contact is a data structure that contains a person's contact details. The application has three activities, **MainActivity**, **AddContactActivity**, and **ShowContactActivity**.

The main activity shows the list of contacts and is shown in Figure 18.1.

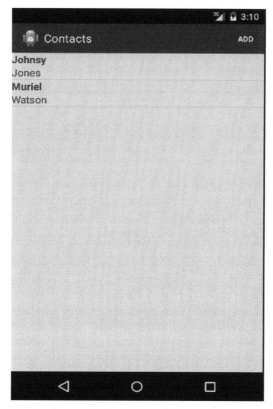

Figure 18.1: The main activity

The main activity offers an Add button on its action bar that will start the
AddContactActivity activity if pressed. The latter activity contains a form for adding a
new contact and is shown in Figure 18.2.

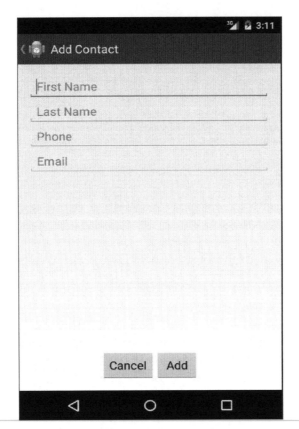

Figure 18.2: AddContactActivity

The main activity also uses a **ListView** to display all contacts in the database. Pressing an item on the list activates the **ShowContactActivity** activity, which is shown in Figure 18.3.

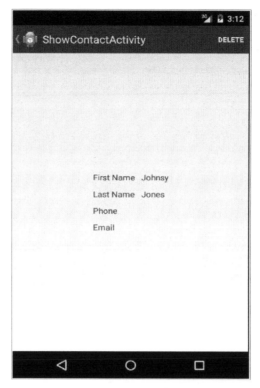

Figure 18.3: ShowContactActivity

The **ShowContactActivity** activity allows the user to delete the shown contact by pressing the Delete button on the action bar. Pressing the button prompts the user to confirm if he or she really wishes to delete the contact. The user can press the activity label to go back to the main activity.

The three activities in the application are declared in the manifest presented in Listing 18.1.

Listing 18.1: The AndroidManifest.xml file

```
<?xml version="1.0" encoding="utf-8"?>
<manifest xmlns:android="http://schemas.android.com/apk/res/android"
    package="com.example.databasedemo1"
    android:versionCode="1"
    android:versionName="1.0" >

    <uses-sdk
        android:minSdkVersion="11"
        android:targetSdkVersion="18" />

    <application
        android:allowBackup="true"
        android:icon="@drawable/ic launcher"
        android:label="@string/app name"
        android:theme="@style/AppTheme" >
        <activity
```

```
                android:name=".MainActivity"
                android:label="@string/app_name" >
                <intent-filter>
                    <action android:name="android.intent.action.MAIN" />
                    <category
android:name="android.intent.category.LAUNCHER"/>
                </intent-filter>
        </activity>
        <activity android:name=".AddContactActivity"
            android:parentActivityName=".MainActivity"
            android:label="@string/title_activity_add_contact">
        </activity>
        <activity android:name=".ShowContactActivity"
            android:parentActivityName=".MainActivity"
            android:label="@string/title_activity_show_contact" >
        </activity>
    </application>
</manifest>
```

DatabaseDemo1 is a simple application that features one object model, the **Contact** class in Listing 18.2. This is a POJO with five properties, **id**, **firstName**, **lastName**, **phone**, and **email**.

Listing 18.2: The Contact class

```
package com.example.databasedemo1;
public class Contact {
    private long id;
    private String firstName;
    private String lastName;
    private String phone;
    private String email;

    public Contact() {
    }

    public Contact(String firstName, String lastName,
            String phone, String email) {
        this.firstName = firstName;
        this.lastName = lastName;
        this.phone = phone;
        this.email = email;
    }
    // get and set methods not shown to save space
}
```

Now comes the most important class in this application, the **DatabaseManager** class in Listing 18.3. This class encapsulates methods for accessing data in the database. The class extends **SQLiteOpenHelper** and implements its **onCreate** and **onUpdate** methods and provides methods for managing contacts, **addContact**, **deleteContact**, **updateContact**, **getAllContacts**, and **getContact**.

Listing 18.3: The DatabaseManager class

```
package com.example.databasedemo1;
```

```java
import java.util.ArrayList;
import java.util.List;
import android.content.ContentValues;
import android.content.Context;
import android.database.Cursor;
import android.database.sqlite.SQLiteDatabase;
import android.database.sqlite.SQLiteOpenHelper;
import android.util.Log;

public class DatabaseManager extends SQLiteOpenHelper {
    public static final String TABLE_NAME = "contacts";
    public static final String ID_FIELD = "_id";
    public static final String FIRST_NAME_FIELD = "first_name";
    public static final String LAST_NAME_FIELD = "last_name";
    public static final String PHONE_FIELD = "phone";
    public static final String EMAIL_FIELD = "email";
    public DatabaseManager(Context context) {
        super(context,
                /*db name=*/ "contacts_db2",
                /*cursorFactory=*/ null,
                /*db version=*/1);
    }
    @Override
    public void onCreate(SQLiteDatabase db) {
        Log.d("db", "onCreate");
        String sql = "CREATE TABLE " + TABLE_NAME
                + " (" + ID_FIELD + " INTEGER, "
                + FIRST_NAME_FIELD + " TEXT,"
                + LAST_NAME_FIELD + " TEXT,"
                + PHONE_FIELD + " TEXT,"
                + EMAIL_FIELD + " TEXT,"
                + " PRIMARY KEY (" + ID_FIELD + "));";
        db.execSQL(sql);

    }


    @Override
    public void onUpgrade(SQLiteDatabase db, int arg1, int arg2) {
        Log.d("db", "onUpdate");
        db.execSQL("DROP TABLE IF EXISTS " + TABLE_NAME);
        // re-create the table
        onCreate(db);
    }

    public Contact addContact(Contact contact) {
        Log.d("db", "addContact");
        SQLiteDatabase db = this.getWritableDatabase();
        ContentValues values = new ContentValues();
        values.put(FIRST_NAME_FIELD, contact.getFirstName());
        values.put(LAST_NAME_FIELD, contact.getLastName());
        values.put(PHONE_FIELD, contact.getPhone());
        values.put(EMAIL_FIELD, contact.getEmail());
        long id = db.insert(TABLE_NAME, null, values);
```

```java
        contact.setId(id);
        db.close();
        return contact;
    }

    // Getting single contact
    Contact getContact(long id) {
        SQLiteDatabase db = this.getReadableDatabase();
        Cursor cursor = db.query(TABLE_NAME, new String[] {
                ID_FIELD, FIRST_NAME_FIELD, LAST_NAME_FIELD,
                PHONE_FIELD, EMAIL_FIELD }, ID_FIELD + "=?",
                new String[] { String.valueOf(id) }, null,
                null, null, null);
        if (cursor != null) {
            cursor.moveToFirst();
            Contact contact = new Contact(
                    cursor.getString(1),
                    cursor.getString(2),
                    cursor.getString(3),
                    cursor.getString(4));
            contact.setId(cursor.getLong(0));
            return contact;
        }
        return null;
    }

    // Getting All Contacts
    public List<Contact> getAllContacts() {
        List<Contact> contacts = new ArrayList<Contact>();
        String selectQuery = "SELECT  * FROM " + TABLE_NAME;

        SQLiteDatabase db = this.getWritableDatabase();
        Cursor cursor = db.rawQuery(selectQuery, null);

        while (cursor.moveToNext()) {
            Contact contact = new Contact();
            contact.setId(Integer.parseInt(cursor.getString(0)));
            contact.setFirstName(cursor.getString(1));
            contact.setLastName(cursor.getString(2));
            contact.setPhone(cursor.getString(3));
            contact.setEmail(cursor.getString(4));
            contacts.add(contact);
        }
        return contacts;
    }

    public Cursor getContactsCursor() {
        String selectQuery = "SELECT  * FROM " + TABLE_NAME;
        SQLiteDatabase db = this.getWritableDatabase();
        return db.rawQuery(selectQuery, null);
    }

    public int updateContact(Contact contact) {
```

```
    SQLiteDatabase db = this.getWritableDatabase();

    ContentValues values = new ContentValues();
    values.put(FIRST_NAME_FIELD, contact.getFirstName());
    values.put(LAST_NAME_FIELD, contact.getLastName());
    values.put(PHONE_FIELD, contact.getPhone());
    values.put(EMAIL_FIELD, contact.getEmail());

    return db.update(TABLE_NAME, values, ID_FIELD + " = ?",
            new String[] { String.valueOf(contact.getId()) });
}

public void deleteContact(long id) {
    SQLiteDatabase db = this.getWritableDatabase();
    db.delete(TABLE_NAME, ID_FIELD + " = ?",
            new String[] { String.valueOf(id) });
    db.close();
}
}
```

The **DatabaseManager** class is used by all the three activity classes. The **MainActivity** class employs a **ListView** that gets its data and layout from a **ListAdapter** that in turn gets its data from a cursor. The **AddContactActivity** class receives the details of a new contact and inserts it into the database by calling the **DatabaseManager** class's **addContact** method. The **ShowContactActivity** class retrieves the details of the pressed contact item in the main activity and uses the **DatabaseManager** class's **getContact** method to achieve this. If the user decides to delete the shown contact, **ShowContactActivity** will resort to **DatabaseManager** to delete it.

The **MainActivity**, **AddContactActivity**, and **ShowContactActivity** classes are given in Listing 18.4, Listing 18.5, and Listing 18.6, respectively.

Listing 18.4: The MainActivity class

```
package com.example.databasedemo1;
import android.app.Activity;
import android.content.Intent;
import android.database.Cursor;
import android.os.Bundle;
import android.support.v4.widget.CursorAdapter;
import android.view.Menu;
import android.view.MenuItem;
import android.view.View;
import android.widget.AdapterView;
import android.widget.AdapterView.OnItemClickListener;
import android.widget.ListAdapter;
import android.widget.ListView;
import android.widget.SimpleCursorAdapter;

public class MainActivity extends Activity {
    DatabaseManager dbMgr;

    @Override
    protected void onCreate(Bundle savedInstanceState) {
```

```
super.onCreate(savedInstanceState);
setContentView(R.layout.activity_main);
ListView listView = (ListView) findViewById(
        R.id.listView);
dbMgr = new DatabaseManager(this);

Cursor cursor = dbMgr.getContactsCursor();
startManagingCursor(cursor);

ListAdapter adapter = new SimpleCursorAdapter(
    this,
    android.R.layout.two_line_list_item,
    cursor,
    new String[] {DatabaseManager.FIRST_NAME_FIELD,
            DatabaseManager.LAST_NAME_FIELD},
    new int[] {android.R.id.text1, android.R.id.text2},
    CursorAdapter.FLAG_REGISTER_CONTENT_OBSERVER);

listView.setAdapter(adapter);
listView.setChoiceMode(ListView.CHOICE_MODE_SINGLE);
listView.setOnItemClickListener(
        new OnItemClickListener() {
    @Override
    public void onItemClick(AdapterView<?> adapterView,
            View view, int position, long id) {
        Intent intent = new Intent(
                getApplicationContext(),
                ShowContactActivity.class);
        intent.putExtra("id", id);
        startActivity(intent);
    }
});
}

@Override
public boolean onCreateOptionsMenu(Menu menu) {
    getMenuInflater().inflate(R.menu.menu_main, menu);
    return true;
}

@Override
public boolean onOptionsItemSelected(MenuItem item) {
    switch (item.getItemId()) {
        case R.id.action_add:
            startActivity(new Intent(this,
                    AddContactActivity.class));
            return true;
        default:
            return super.onOptionsItemSelected(item);
    }
}
}
```

Listing 18.5: The AddContactActivity class

```
package com.example.databasedemo1;
import android.app.Activity;
import android.os.Bundle;
import android.view.Menu;
import android.view.View;
import android.widget.TextView;

public class AddContactActivity extends Activity {

    @Override
    protected void onCreate(Bundle savedInstanceState) {
        super.onCreate(savedInstanceState);
        setContentView(R.layout.activity_add_contact);
    }

    @Override
    public boolean onCreateOptionsMenu(Menu menu) {
        getMenuInflater().inflate(R.menu.add_contact, menu);
        return true;
    }

    public void cancel(View view) {
        finish();
    }
    public void addContact(View view) {
        DatabaseManager dbMgr = new DatabaseManager(this);
        String firstName = ((TextView) findViewById(
                R.id.firstName)).getText().toString();
        String lastName = ((TextView) findViewById(
                R.id.lastName)).getText().toString();
        String phone = ((TextView) findViewById(
                R.id.phone)).getText().toString();
        String email = ((TextView) findViewById(
                R.id.email)).getText().toString();
        Contact contact = new Contact(firstName, lastName,
                phone, email);
        dbMgr.addContact(contact);
        finish();
    }
}
```

Listing 18.6: The ShowContactActivity class

```
package com.example.databasedemo1;
import android.app.Activity;
import android.app.AlertDialog;
import android.content.DialogInterface;
import android.os.Bundle;
import android.util.Log;
import android.view.Menu;
import android.view.MenuItem;
import android.widget.TextView;
```

```java
public class ShowContactActivity extends Activity {
    long contactId;

    @Override
    protected void onCreate(Bundle savedInstanceState) {
        super.onCreate(savedInstanceState);
        setContentView(R.layout.activity_show_contact);
        getActionBar().setDisplayHomeAsUpEnabled(true);
        Bundle extras = getIntent().getExtras();
        if (extras != null) {
            contactId = extras.getLong("id");
            DatabaseManager dbMgr = new DatabaseManager(this);
            Contact contact = dbMgr.getContact(contactId);
            if (contact != null) {
                ((TextView) findViewById(R.id.firstName))
                        .setText(contact.getFirstName());
                ((TextView) findViewById(R.id.lastName))
                        .setText(contact.getLastName());
                ((TextView) findViewById(R.id.phone))
                        .setText(contact.getPhone());
                ((TextView) findViewById(R.id.email))
                        .setText(contact.getEmail());
            } else {
                Log.d("db", "contact null");
            }
        }
    }

    @Override
    public boolean onCreateOptionsMenu(Menu menu) {
        getMenuInflater().inflate(R.menu.show_contact, menu);
        return true;
    }

    @Override
    public boolean onOptionsItemSelected(MenuItem item) {
        switch (item.getItemId()) {
        case R.id.action_delete:
            deleteContact();
            return true;
        default:
            return super.onOptionsItemSelected(item);
        }
    }

    private void deleteContact() {
        new AlertDialog.Builder(this)
            .setTitle("Please confirm")
            .setMessage(
                "Are you sure you want to delete " +
                "this contact?")
            .setPositiveButton("Yes",
                new DialogInterface.OnClickListener() {
```

```
        public void onClick(
                DialogInterface dialog,
                int whichButton) {
            DatabaseManager dbMgr =
                    new DatabaseManager(
                    getApplicationContext());
            dbMgr.deleteContact(contactId);
            dialog.dismiss();
            finish();
        }
    })
    .setNegativeButton("No",
        new DialogInterface.OnClickListener() {
            public void onClick(
                    DialogInterface dialog,
                    int which) {
                dialog.dismiss();
            }
        })
    .create()
    .show();
    }
}
```

Summary

The Android Database API makes it easy to work with relational databases. The
android.database and **android.database.sqlite** packages contains classes and interfaces
that support access to a SQLite database, which is the default database shipped with
Android. In this chapter you learned how to use the three most frequently used types in
the API, the **SQLiteOpenHelper** class, the **SQLiteDatabase** class, and the **Cursor**
interface.

Chapter 19
Taking Pictures

Almost all Android handsets and tablets come with one or two cameras. You can use a camera to take still pictures by starting an activity in the built-in Camera application or use the Camera API.

This chapter shows how to use both approaches.

Overview

An Android application can call another application to use one or two features offered by the latter. For example, to send an email from your application, you can use the default Email application rather than writing your own app. In the case of taking a picture, the easiest way to do this is by using the Camera application. To activate Camera, use the following code.

```
int requestCode = ...;
Intent intent = new Intent(MediaStore.ACTION_IMAGE_CAPTURE);
startActivityForResult(intent, requestCode);
```

Basically, you need to create an **Intent** by passing **MediaStore.ACTION_IMAGE_CAPTURE** to the **Intent** class's constructor. Then, you need to call **startActivityForResult** from your activity passing the **Intent** and a request code. The request code can be any integer your heart desires. You will learn shortly the purpose of passing a request code.

To tell Camera where to save the taken picture, you can pass a **Uri** to the **Intent**. Here is the complete code.

```
int requestCode = ...;
Intent intent = new Intent(MediaStore.ACTION_IMAGE_CAPTURE);
Uri uri = ...;
intent.putExtra(MediaStore.EXTRA_OUTPUT, uri);
startActivityForResult(intent, requestCode);
```

When the user closes Camera after taking a picture or canceling the operation, Android will notify your application by calling the **onActivityResult** method in the activity that called Camera. This gives you the opportunity to save the picture taken using Camera. The signature of **onActivityResult** is as follows.

```
protected void onActivityResult(int requestCode, int resultCode,
        android.content.Intent data)
```

The system calls **onActivityResult** by passing three arguments. The first argument, **requestCode**, is the request code passed when calling **startActivityForResult**. The request code is important if you call other activities from your activity, passing a different request code each time. Since you can only have one **onActivityResult** implementation in your activity, all calls to **startActivityForResult** will share the same **onActivityResult**, and you need to know which activity caused **onActivityResult** to be called by checking the request code.

The second argument to **onActivityResult** is a result code. The value can be either **Activity.RESULT_OK** or **Activity.RESULT_CANCELED** or a user defined value. **Activity.RESULT_OK** indicates that the operation succeeded and **Activity.RESULT_CANCELED** indicates that the operation was canceled.

The third argument to **onActivityResult** contains data from the called activity if the operation was successful.

Using Camera is easy. However, if Camera does not suit your needs, you can also use the Camera API directly. This is not as easy as using Camera, but the API lets you configure many aspects of the camera.

The samples accompanying this chapter show you both methods.

Using Camera

To use the camera, you need these in your manifest.

```
<uses-feature android:name="android.hardware.camera"/>
<uses-permission android:name="android.permission.CAMERA"/>
```

The CameraDemo application shows how to use the built-in intent to activate the Camera application and use it to take a picture. CameraDemo has only one activity, which sports two button on its action bar, Show Camera and Email. The Show Camera button starts Camera and the Email button emails the picture. The application is shown in Figure 19.1.

Figure 19.1: CameraDemo

Let's start dissecting the code, starting from the manifest.

Listing 19.1: The manifest

```xml
<?xml version="1.0" encoding="utf-8"?>
<manifest xmlns:android="http://schemas.android.com/apk/res/android"
    package="com.example.camerademo" >
    <uses-feature android:name="android.hardware.camera"/>
    <uses-permission android:name="android.permission.CAMERA"/>
    <uses-permission
        android:name="android.permission.WRITE_EXTERNAL_STORAGE"/>

    <application
        android:allowBackup="true"
        android:icon="@drawable/ic_launcher"
        android:label="@string/app_name"
        android:theme="@style/AppTheme" >
        <activity
            android:name="com.example.camerademo.MainActivity"
            android:label="@string/app_name" >
            <intent-filter>
                <action android:name="android.intent.action.MAIN" />

                <category android:name="android.intent.category.LAUNCHER"
    />
            </intent-filter>
        </activity>
    </application>

</manifest>
```

menu file (the **main.xml** file in Listing 19.1) that contains two menu items for the action bar.

Listing 19.1: The menu file (menu_main.xml)

```xml
<menu xmlns:android="http://schemas.android.com/apk/res/android" >

    <item
        android:id="@+id/action_camera"
        android:orderInCategory="100"
        android:showAsAction="ifRoom"
        android:title="@string/action_show_camera"/>
    <item
        android:id="@+id/action_email"
        android:orderInCategory="200"
        android:showAsAction="ifRoom"
        android:title="@string/action_email"/>
</menu>
```

The layout file for the main activity is presented in Listing 19.2. It contains an **ImageView** for showing the taken picture. The activity class itself is shown in Listing 19.3.

Listing 19.2: The activity_main.xml file

```xml
<RelativeLayout
        xmlns:android="http://schemas.android.com/apk/res/android"
    android:layout_width="match_parent"
    android:layout_height="match_parent"
    android:paddingBottom="@dimen/activity_vertical_margin"
    android:paddingLeft="@dimen/activity_horizontal_margin"
    android:paddingRight="@dimen/activity_horizontal_margin"
    android:paddingTop="@dimen/activity_vertical_margin">

    <ImageView
        android:id="@+id/imageView"
        android:layout_width="match_parent"
        android:layout_height="match_parent"
    />
</RelativeLayout>
```

Listing 19.3: The MainActivity class

```java
package com.example.camerademo;
import java.io.File;
import android.app.Activity;
import android.content.Intent;
import android.net.Uri;
import android.os.Bundle;
import android.os.Environment;
import android.provider.MediaStore;
import android.util.Log;
import android.view.Menu;
import android.view.MenuItem;
import android.widget.ImageView;
import android.widget.Toast;
```

```java
public class MainActivity extends Activity {
    private static final int CAPTURE_IMAGE_ACTIVITY_REQUEST_CODE = 100;
    File pictureDir = new
        File(Environment.getExternalStoragePublicDirectory(
            Environment.DIRECTORY_PICTURES), "CameraDemo");
    private static final String FILE_NAME = "image01.jpg";

    private Uri fileUri;

    @Override
    protected void onCreate(Bundle savedInstanceState) {
        super.onCreate(savedInstanceState);
        setContentView(R.layout.activity_main);
        if (!pictureDir.exists()) {
            pictureDir.mkdirs();
        }
    }

    @Override
    public boolean onCreateOptionsMenu(Menu menu) {
        getMenuInflater().inflate(R.menu.menu_main, menu);
        return true;
    }

    @Override
    public boolean onOptionsItemSelected(MenuItem item) {
        switch (item.getItemId()) {
            case R.id.action_camera:
                showCamera();
                return true;
            case R.id.action_email:
                emailPicture();
                return true;
            default:
                return super.onContextItemSelected(item);
        }
    }

    private void showCamera() {
        Intent intent = new Intent(
                MediaStore.ACTION_IMAGE_CAPTURE);
        File image = new File(pictureDir, FILE_NAME);
        fileUri = Uri.fromFile(image);
        intent.putExtra(MediaStore.EXTRA_OUTPUT, fileUri);
        // check if the device has a camera:
        if (intent.resolveActivity(getPackageManager()) != null) {
            startActivityForResult(intent,
                    CAPTURE_IMAGE_ACTIVITY_REQUEST_CODE);
        }
    }

    @Override
```

```
protected void onActivityResult(int requestCode,
        int resultCode, Intent data) {
    if (requestCode ==
            CAPTURE_IMAGE_ACTIVITY_REQUEST_CODE) {
        if (resultCode == RESULT_OK) {
            ImageView imageView = (ImageView)
                    findViewById(R.id.imageView);
            File image = new File(pictureDir, FILE_NAME);
            fileUri = Uri.fromFile(image);
            imageView.setImageURI(fileUri);
        } else if (resultCode == RESULT_CANCELED) {
            Toast.makeText(this, "Action cancelled",
                    Toast.LENGTH_LONG).show();
        } else {
            Toast.makeText(this, "Error",
                    Toast.LENGTH_LONG).show();
        }
    }
}

private void emailPicture() {
    Intent emailIntent = new Intent(
            android.content.Intent.ACTION_SEND);
    emailIntent.setType("application/image");
    emailIntent.putExtra(android.content.Intent.EXTRA_EMAIL,
            new String[]{"me@example.com"});
    emailIntent.putExtra(android.content.Intent.EXTRA_SUBJECT,
            "New photo");
    emailIntent.putExtra(android.content.Intent.EXTRA_TEXT,
            "From My App");
    emailIntent.putExtra(Intent.EXTRA_STREAM, fileUri);
    startActivity(Intent.createChooser(emailIntent,
            "Send mail..."));
}
}
```

The Show Camera button in **MainActivity** calls the **showCamera** method. This method starts Camera by calling **startActivityForResult**. The **emailPicture** method starts another activity that in turn activates the default Email application.

The Camera API

At the center of the Camera API is the **android.hardware.Camera** class. A **Camera** represents a digital camera.

Every camera has a viewfinder, through which the photographer can see what the camera is seeing. A viewfinder can be optical or electronic. An analog camera normally offers an optical viewfinder, which is a reversed telescope mounted on the camera body. Some digital cameras have an electronic viewfinder and some have an electronic one plus an optical one. On an Android tablet and handset, the whole screen or part of the screen is normally used as a viewfinder.

In an application that uses a camera, the **android.view.SurfaceView** class is normally used as a viewfinder. **SurfaceView** is a subclass of **View** and, as such, can be added to an activity by declaring it in a layout file using the **SurfaceView** element. The area of a **SurfaceView** will be continuously updated with what the camera sees. You control a **SurfaceView** through its **SurfaceHolder**, which you can obtain by calling the **getHolder** method on the **SurfaceView**. **SurfaceHolder** is an interface in the **android.view** package.

Therefore, when working with a camera, you need to manage an instance of **Camera** as well as a **SurfaceHolder**.

Managing A Camera

When working with the Camera API, you should start by checking if the device does have a camera. You must also determine which camera to use if a device has multiple cameras. You do it by calling the **open** static method of the **Camera** class.

```
Camera camera = null;
try {
    if (Build.VERSION.SDK_INT >= Build.VERSION_CODES.GINGERBREAD) {
        camera = Camera.open(0);
    } else {
        camera = Camera.open();
    }
} catch (Exception e) {
    e.printStackTrace();
}
```

For pre-Gingerbread Android (Android version 2.3), use the no-argument method overload. For Android version 2.3, use the overload that takes an integer.

```
public static Camera open(int cameraId)
```

Passing 0 to the method gives you the first camera, 1 the second camera, and so on.

You should enclose the call to **open** in a try block as it may throw an exception.

Once you obtain a **Camera**, pass a **SurfaceHolder** to the **setPreviewDisplay** method on the **Camera**.

```
public void setPreviewDisplay(android.view.SurfaceHolder holder)
```

If **setPreviewDisplay** returns successfully, call the camera's **startPreview** method and the **SurfaceView** controlled by the **SurfaceHolder** will start displaying what the camera sees.

To take a picture, call the camera's **takePicture** method. After a picture is taken, the preview will stop so you will need to call **startPreview** again to take another picture.

When you are finished with the camera, call **stopPreview** and **release** to release the camera.

Optionally, you can configure the camera after you call **open** by calling its **getParameters** method, modifying the parameters, and passing them back to the camera using the **setParameters** method.

With the **takePicture** method you can decide what to do to the resulting raw and JPEG images from the camera. The signature of **takePicture** is as follows.

```
public final void takePicture(Camera.ShutterCallback shutter,
    Camera.PictureCallback raw, Camera.PictureCallback postview,
    Camera.PictureCallback jpeg)
```

The four parameters are these.

- *shutter*. The callback for image capture moment. For example, you can pass code that plays a click sound to make it more like a real camera.
- *raw*. The callback for uncompressed image data.
- *postview*. The callback with postview image data.
- *jpeg*. The callback for JPEG image data.

You will learn how to use **Camera** in the **CameraAPIDemo** application.

Managing A SurfaceHolder

A **SurfaceHolder** communicates with its user through a series of methods in **SurfaceHolder.Callback**. To manage a **SurfaceHolder**, you need to pass an instance of **SurfaceHolder.Callback** to the **SurfaceHolder**'s **addCallback** method.

SurfaceHolder.Callback exposes these three methods that the **SurfaceHolder** will call in response to events.

```
public abstract void surfaceChanged(SurfaceHolder holder,
        int format, int width, int height)
```
Called after any structural changes (format or size) have been made to the surface.

```
public abstract void surfaceCreated(SurfaceHolder holder)
```
Called after the surface is first created.

```
public abstract void surfaceDestroyed(SurfaceHolder holder)
```
Called before a surface is being destroyed.

For instance, you might want to link a **SurfaceHolder** with a **Camera** right after the **SurfaceHolder** is created. Therefore, you might want to override the **surfaceCreated** method with this code.

```
@Override
public void surfaceCreated(SurfaceHolder holder) {
    try {
        camera.setPreviewDisplay(holder);
        camera.startPreview();
    } catch (Exception e){
        Log.d("camera", e.getMessage());
    }
}
```

Using the Camera API

The CameraAPIDemo application demonstrates the use of the Camera API to take still pictures. It uses a **SurfaceView** as a viewfinder and a button to take a picture. Clicking the button takes the picture and emits a beep sound. After a picture is taken, the

SurfaceView freezes for two seconds to give the user the change to inspect the picture and restart the camera preview to allow the user to take another picture. All pictures are given a random name and stored in the external storage.

The application has one activity, whose layout is shown in Listing 19.4.

Listing 19.4: The layout file (activity_main.xml)

```
<LinearLayout xmlns:android="http://schemas.android.com/apk/res/android"
    android:orientation="vertical"
    android:layout_width="fill_parent"
    android:layout_height="fill_parent">

    <Button
        android:id="@+id/button1"
        android:layout_width="wrap_content"
        android:layout_height="wrap_content"
        android:layout_gravity="center"
        android:onClick="takePicture"
        android:text="@string/button_take"/>

    <SurfaceView
        android:id="@+id/surfaceview"
        android:layout_width="match_parent"
        android:layout_height="match_parent" />
</LinearLayout>
```

The layout features a **LinearLayout** containing a button and a **SurfaceView**. The activity class is presented in Listing 19.5.

Listing 19.5: The MainActivity class

```
package com.example.cameraapidemo;
import java.io.File;
import java.io.FileNotFoundException;
import java.io.FileOutputStream;
import java.io.IOException;
import android.app.Activity;
import android.hardware.Camera;
import android.hardware.Camera.PictureCallback;
import android.hardware.Camera.ShutterCallback;
import android.media.AudioManager;
import android.media.SoundPool;
import android.net.Uri;
import android.os.Build;
import android.os.Bundle;
import android.os.Environment;
import android.os.Handler;
import android.provider.Settings;
import android.util.Log;
import android.view.Menu;
import android.view.SurfaceHolder;
import android.view.SurfaceView;
import android.view.View;
import android.widget.Button;
import android.widget.Toast;
```

```java
public class MainActivity extends Activity
        implements SurfaceHolder.Callback {

    private Camera camera;
    SoundPool soundPool;
    int beepId;
    File pictureDir = new File(Environment
            .getExternalStoragePublicDirectory(
                    Environment.DIRECTORY_PICTURES),
                    "CameraAPIDemo");
    private static final String TAG = "camera";

    @Override
    public void onCreate(Bundle savedInstanceState) {
        super.onCreate(savedInstanceState);
        setContentView(R.layout.activity_main);
        pictureDir.mkdirs();

        soundPool = new SoundPool(1,
                AudioManager.STREAM_NOTIFICATION, 0);
        Uri uri = Settings.System.DEFAULT_RINGTONE_URI;
        beepId = soundPool.load(uri.getPath(), 1);
        SurfaceView surfaceView = (SurfaceView)
                findViewById(R.id.surfaceview);
        surfaceView.getHolder().addCallback(this);
    }

    @Override
    public boolean onCreateOptionsMenu(Menu menu) {
        getMenuInflater().inflate(R.menu.menu_main, menu);
        return true;
    }

    @Override
    public void onResume() {
        super.onResume();
        try {
            if (Build.VERSION.SDK_INT >=
                    Build.VERSION_CODES.GINGERBREAD) {
                camera = Camera.open(0);
            } else {
                camera = Camera.open();
            }
        } catch (Exception e) {
            e.printStackTrace();
        }
    }

    @Override
    public void onPause() {
        super.onPause();
        if (camera != null) {
```

```
            try {
                camera.release();
                camera = null;
            } catch (Exception e) {
                e.printStackTrace();
            }
        }
    }

    private void enableButton(boolean enabled) {
        Button button = (Button) findViewById(R.id.button1);
        button.setEnabled(enabled);
    }

    public void takePicture(View view) {
        enableButton(false);
        camera.takePicture(shutterCallback, null,
                pictureCallback);
    }

    private ShutterCallback shutterCallback =
            new ShutterCallback() {
        @Override
        public void onShutter() {
            // play sound
            soundPool.play(beepId, 1.0f, 1.0f, 0, 0, 1.0f);
        }
    };

    private PictureCallback pictureCallback =
            new PictureCallback() {
        @Override
        public void onPictureTaken(byte[] data,
                final Camera camera) {
            Toast.makeText(MainActivity.this, "Saving image",
                    Toast.LENGTH_LONG)
                    .show();
            File pictureFile = new File(pictureDir,
                    System.currentTimeMillis() + ".jpg");

            try {
                FileOutputStream fos = new FileOutputStream(
                        pictureFile);
                fos.write(data);
                fos.close();
            } catch (FileNotFoundException e) {
                Log.d(TAG, e.getMessage());
            } catch (IOException e) {
                Log.d(TAG, e.getMessage());
            }

            Handler handler = new Handler();
            handler.postDelayed(new Runnable() {
```

```
                    @Override
                    public void run() {
                        try {
                            enableButton(true);
                            camera.startPreview();
                        } catch (Exception e) {
                            Log.d("camera",
                                    "Error starting camera preview: "
                                            + e.getMessage());
                        }
                    }
                }, 2000);
            }
        };

        @Override
        public void surfaceCreated(SurfaceHolder holder) {
            try {
                camera.setPreviewDisplay(holder);
                camera.startPreview();
            } catch (Exception e){
                Log.d("camera", e.getMessage());
            }
        }

        @Override
        public void surfaceChanged(SurfaceHolder holder,
                int format, int w, int h3) {
            if (holder.getSurface() == null){
                Log.d(TAG, "surface does not exist, return");
                return;
            }

            try {
                camera.setPreviewDisplay(holder);
                camera.startPreview();
            } catch (Exception e){
                Log.d("camera", e.getMessage());
            }
        }

        @Override
        public void surfaceDestroyed(SurfaceHolder holder) {
            Log.d(TAG, "surfaceDestroyed");
        }
    }
```

The **MainActivity** class uses a **Camera** and a **SurfaceView**. The latter continuously displays what the camera sees. Since a **Camera** takes a lot of resources to operate, the **MainActivity** releases the camera when the application stops and re-opens it when the application resumes.

The **MainActivity** class also implements **SurfaceHolder.Callback** and passes itself to the **SurfaceHolder** of the **SurfaceView** it employs as a viewfinder. This is shown in the following lines in the **onCreate** method.

```
SurfaceView surfaceView = (SurfaceView)
        findViewById(R.id.surfaceview);
surfaceView.getHolder().addCallback(this);
```

In both **surfaceCreated** and **surfaceChanged** methods that **MainActivity** overrides, the class calls the camera's **setPreviewDisplay** and **startPreview** methods. This makes sure when the camera is linked with a **SurfaceHolder**, the **SurfaceHolder** has already been created.

Another important point in **MainActivity** is the **takePicture** method that gets called when the user presses the button.

```
public void takePicture(View view) {
    enableButton(false);
    camera.takePicture(shutterCallback, null,
            pictureCallback);
}
```

The **takePicture** method disables the button so that no more picture can be taken until the picture is saved and calls the **takePicture** method on the **Camera**, passing a **Camera.ShutterCallback** and a **Camera.PictureCallback**. Note that calling **takePicture** on a **Camera** also stops previewing the image on the **SurfaceHolder** linked to the camera.

The **Camera.ShutterCallback** in **MainActivity** has one method, **onShutter**, that plays a sound from the sound pool.

```
@Override
public void onShutter() {
    // play sound
    soundPool.play(beepId, 1.0f, 1.0f, 0, 0, 1.0f);
}
```

The **Camera.PictureCallback** also has one method, **onPictureTaken**, whose signature is this.

```
public void onPictureTaken(byte[] data, final Camera camera)
```

This method is called by the **Camera**'s **takePicture** method and receives a byte array containing the photo image.

The **onPictureTaken** method implementation in **MainActivity** does three things. First, it displays a message using the **Toast**. Second, it saves the byte array into a file. The name of the file is generated using **System.currentTimeMillis()**. Finally, the method creates a **Handler** to schedule a task that will be executed in two seconds. The task enables the button and calls the camera's **startPreview** so that the viewfinder will start working again.

Figure 19.2 shows the CameraAPIDemo application.

Figure 19.2: The CameraAPIDemo application

Summary

Android offers two options for applications that need to take still pictures: use a built-in intent to start Camera or use the Camera API. The first option is the easier one to use but lacks the features that the Camera API provides.

This chapter showed how to use both methods.

Chapter 20
Making Videos

The easiest way to provide video-making capability in your application is to use a built-in intent to activate an existing activity. However, if you need more than what the default application can provide, you need to get your hands dirty and work with the API directly.

This chapter shows how to use both methods for making videos.

Using the Built-in Intent

If you choose to use the default Camera application for making video, you can activate the application with these three lines of code.

```
int requestCode = ...;
Intent intent = new Intent(MediaStore.ACTION_VIDEO_CAPTURE);
startActivityForResult(intent, requestCode);
```

Basically, you need to create an **Intent** object by passing **MediaStore.ACTION_VIDEO_CAPTURE** to its constructor and pass it to the **startActivityForResult** method in your activity class. You can choose any integer for the request code that you will pass as the second argument to **startActivityForResult**. This method will pause the current activity and start Camera and make it ready to capture a video.

When you exit from Camera, either by canceling the operation or when you are done making a video, the system will resume your original activity (the activity where you called **startActivityForResult**) and call its **onActivityResult** method. If you are interested in saving or processing the captured video, you must override **onActivityResult**. Its signature is as follows.

```
protected void onActivityResult(int requestCode, int resultCode,
        android.content.Intent data)
```

The system calls **onActivityResult** by passing three arguments. The first argument, **requestCode**, is the request code passed when you called **startActivityForResult**. The request code is important if you are calling other activities from your activity, passing a different request code for each activity. Since you can only have one **onActivityResult** implementation in your activity, all calls to **startActivityForResult** will share the same **onActivityResult**, and you need to know which activity caused **onActivityResult** to be called by checking the value of the request code.

The second argument to **onActivityResult** is a result code. The value can be either **Activity.RESULT_OK** or **Activity.RESULT_CANCELED** or a user defined value. **Activity.RESULT_OK** indicates that the operation succeeded and **Activity.RESULT_CANCELED** indicates that the operation was canceled.

The third argument to **onActivityResult** contains data from Camera if the operation was successful.

In addition, you need the following **uses-feature** element in your manifest to indicate that your application needs to use the camera hardware of the device.

```
<uses-feature android:name="android.hardware.camera"
        android:required="true" />
```

As an example, consider the VideoDemo application that has an activity with a button on its action bar. You can press this button to activate Camera for the purpose of making a video. The VideoDemo application is shown in Figure 20.1.

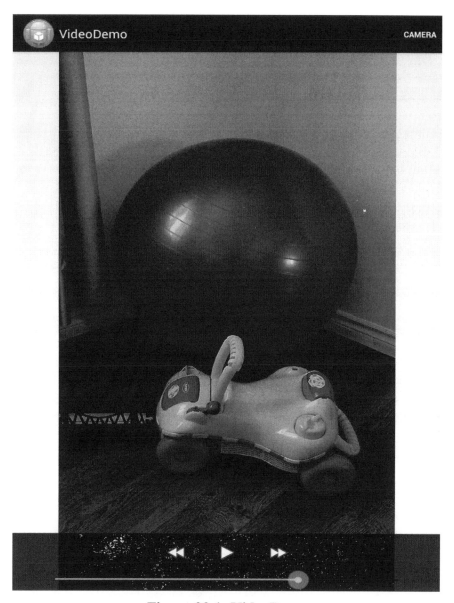

Figure 20.1: VideoDemo

The **AndroidManifest.xml** file in Listing 20.1 shows the activity used in the application as well as a **use-feature** element.

Listing 20.1: The AndroidManifest.xml file

```
<?xml version="1.0" encoding="utf-8"?>
<manifest xmlns:android="http://schemas.android.com/apk/res/android"
    package="com.example.videodemo"
    android:versionCode="1"
    android:versionName="1.0" >
```

```
<uses-sdk
    android:minSdkVersion="16"
    android:targetSdkVersion="19" />

<uses-feature android:name="android.hardware.camera"
              android:required="true" />

<application
    android:allowBackup="true"
    android:icon="@drawable/ic_launcher"
    android:label="@string/app_name"
    android:theme="@style/AppTheme" >
    <activity
        android:name="com.example.videodemo.MainActivity"
        android:label="@string/app_name">
        <intent-filter>
            <action android:name="android.intent.action.MAIN" />
            <category android:name="android.intent.category.LAUNCHER"
    />
        </intent-filter>
    </activity>
</application>
</manifest>
```

There is only one activity in the application, the **MainActivity** activity. The activity reads the menu file in Listing 20.2 to populate its action bar.

Listing 20.2: The menu file (menu_main.xml)

```
<menu xmlns:android="http://schemas.android.com/apk/res/android">
    <item
        android:id="@+id/action_camera"
        android:orderInCategory="100"
        android:showAsAction="always"
        android:title="@string/action_camera"/>
</menu>
```

The activity also uses the layout file in Listing 20.3 to set its view. There is only a **FrameLayout** with a **VideoView** element that is used to display the video file.

Listing 20.3: The activity_main.xml file

```
<FrameLayout xmlns:android="http://schemas.android.com/apk/res/android"
    android:layout_width="match_parent"
    android:layout_height="match_parent" >
    <VideoView
        android:id="@+id/videoView"
        android:layout_width="match_parent"
        android:layout_height="match_parent"
        android:layout_gravity="center">
    </VideoView>
</FrameLayout>
```

Note that I use a **FrameLayout** to enclose the **VideoView** to center it. For some reason, a **LinearLayout** or a **RelativeLayout** will not center it.

Finally, the **MainActivity** class is presented in Listing 20.4. You should know by now that the **onOptionsItemSelected** method is called when the a menu item is pressed. In short, pressing the Camera button on the action bar calls the **showCamera** method. **showCamera** constructs a built-in **Intent** and passes it to **startActivityForResult** to activate the video making feature in Camera.

Listing 20.4: The MainActivity class

```
package com.example.videodemo;
import android.app.Activity;
import android.content.Intent;
import android.net.Uri;
import android.os.Bundle;
import android.provider.MediaStore;
import android.view.Menu;
import android.view.MenuItem;
import android.widget.MediaController;
import android.widget.Toast;
import android.widget.VideoView;

public class MainActivity extends Activity {
    private static final int REQUEST_CODE = 200;

    @Override
    protected void onCreate(Bundle savedInstanceState) {
        super.onCreate(savedInstanceState);
        setContentView(R.layout.activity_main);
    }

    @Override
    public boolean onCreateOptionsMenu(Menu menu) {
        getMenuInflater().inflate(R.menu.menu_main, menu);
        return true;
    }

    @Override
    public boolean onOptionsItemSelected(MenuItem item) {
        switch (item.getItemId()) {
        case R.id.action_camera:
            showCamera();
            return true;
        default:
            return super.onContextItemSelected(item);
        }
    }

    private void showCamera() {
        // cannot set the video file
        Intent intent = new Intent(
                MediaStore.ACTION_VIDEO_CAPTURE);
        // check if the device has a camera:
        if (intent.resolveActivity(getPackageManager()) != null) {
            startActivityForResult(intent, REQUEST_CODE);
        } else {
```

```
                Toast.makeText(this, "Opening camera failed",
                        Toast.LENGTH_LONG).show();
            }
        }

    @Override
    protected void onActivityResult(int requestCode,
            int resultCode, Intent data) {
        if (requestCode == REQUEST_CODE) {
            if (resultCode == RESULT_OK) {
                if (data != null) {
                    Uri uri = data.getData();
                    VideoView videoView = (VideoView)
                            findViewById(R.id.videoView);

                    videoView.setVideoURI(uri);
                    videoView.setMediaController(
                            new MediaController(this));
                    videoView.requestFocus();
                }
            } else if (resultCode == RESULT_CANCELED) {
                Toast.makeText(this, "Action cancelled",
                        Toast.LENGTH_LONG).show();
            } else {
                Toast.makeText(this, "Error", Toast.LENGTH_LONG)
                        .show();
            }
        }
    }
}
```

What is interesting is the implementation of the **onActivityResult** method, which gets called when the user leaves Camera. If the result code is **RESULT_OK** and **data** is not null, the method calls the **getData** method on **data** to get a **Uri** pointing to the location of the video. Next, it finds the **VideoView** widget and set its **videoURI** property and calls two other methods on the **VideoView**, **setMediaController** and **requestFocus**.

```
    protected void onActivityResult(int requestCode,
            int resultCode, Intent data) {
        if (requestCode == REQUEST_CODE) {
            if (resultCode == RESULT_OK) {
                if (data != null) {
                    Uri uri = data.getData();
                    VideoView videoView = (VideoView)
                            findViewById(R.id.videoView);

                    videoView.setVideoURI(uri);
                    videoView.setMediaController(
                            new MediaController(this));
                    videoView.requestFocus();
                }
    ...
```

Passing a **MediaController** decorates the **VideoView** with a media controller that can be used to play and stop the video. Calling **requestFocus()** on the VideoView sets focus on the widget.

MediaRecorder

If you choose to deal with the API directly rather than using the Camera to provide your application with video-making capability, you need to know the details of **MediaRecorder**.

The **android.media.MediaRecorder** class can be used to record audio and video. Figure 20.2 shows the various states a **MediaRecord** can be in.

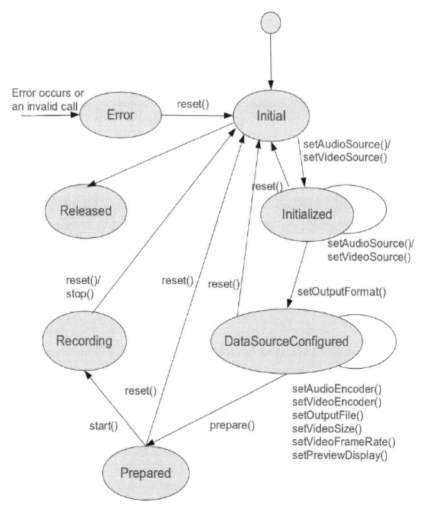

Figure 20.2: The MediaRecorder state diagram

To capture a video with a **MediaRecorder**, of course you need an instance of it. So, the first thing to do is to create a **MediaRecorder**.

```
MediaRecorder mediaRecorder = new MediaRecorder();
```

Then, as you can see in Figure 20.2, to record a video, you have to bring the **MediaRecorder** to the **Initialized** state, followed by the **DataSourceConfigured** and **Prepared** states by calling certain methods.

To transition a **MediaRecorder** to the **Initialized** state, call the **setAudioSource** and **setVideoSource** methods to set the audio and video sources. The valid value for **setAudioSource** is one of the fields defined in the **MediaRecorder.AudioSource** class, which are **CAMCORDER**, **DEFAULT**, **MIC**, **REMOTE_SUBMIX**, **VOICE_CALL**, **VOICE_COMMUNICATION**, **VOICE_DOWNLINK**, **VOICE_RECOGNITION**, and **VOICE_UPLINK**.

The valid value for **setVideoSource** is one of the fields in the **MediaRecorder.VideoSource** class, which are **CAMERA** and **DEFAULT**.

Once the **MediaRecorder** is in the **Initialized** state, call its **setOutputFormat** method, passing one of the file formats in the **MediaRecorder.OutputFormat** class. The following fields are defined: **AAC_ADTS**, **AMR_NB**, **AMR_WB**, **DEFAULT**, **MPEG_4**, **RAW_AMR**, and **THREE_GPP**.

Successfully calling **setOutputFormat** brings the **MediaRecorder** to the **DataSourceConfigured** state. You just need to call **prepare** to prepare the **MediaRecorder**.

To start recording, call the **start** method. It will keep recording until **stop** is called or an error occurs. An error may occur if the **MediaRecorder** runs out of space to store video or if a specified maximum record time is exceeded.

Once you stop a **MediaRecorder**, it goes back to the initial state. You must take it through the previous three states again to record another video.

Also note that a **MediaRecorder** uses a lot of resources and it is prudent to release the resources by calling its **release** method if the **MediaRecorder** is not being used. For example, you should release the **MediaRecorder** when the activity is paused. Once a **MediaRecorder** is released, the same instance cannot be reused to record another video.

Using MediaRecorder

The **VideoRecorder** application demonstrates how to use the **MediaRecorder** to record a video. It has one activity that contains a button and a **SurfaceView**. The button is used to start and stop recording whereas the **SurfaceView** for displaying what the camera sees. **SurfaceView** was explained in detail in Chapter 19, "Taking Pictures."

The layout file for the activity is shown in Listing 20.5 and the activity class in Listing 20.6.

Listing 20.5: The layout file (activity_main.xml)

```
<LinearLayout xmlns:android="http://schemas.android.com/apk/res/android"
    android:orientation="vertical"
    android:layout_width="fill_parent"
    android:layout_height="fill_parent">
```

```
    <Button
        android:id="@+id/button1"
        android:layout_width="wrap_content"
        android:layout_height="wrap_content"
        android:layout_marginLeft="33dp"
        android:layout_marginTop="22dp"
        android:onClick="startStopRecording"
        android:text="@string/button_start" />
    <SurfaceView
        android:id="@+id/surfaceView"
        android:layout_width="match_parent"
        android:layout_height="match_parent" />
</LinearLayout>
```

Listing 20.6: The MainActivity class

```java
package com.example.videorecorder;
import java.io.File;
import java.io.IOException;
import android.app.Activity;
import android.media.MediaRecorder;
import android.os.Bundle;
import android.os.Environment;
import android.view.SurfaceHolder;
import android.view.SurfaceView;
import android.view.View;
import android.widget.Button;

public class MainActivity extends Activity {
    private MediaRecorder mediaRecorder;
    private File outputDir;
    private boolean recording = false;

    @Override
    public void onCreate(Bundle savedInstanceState) {
        super.onCreate(savedInstanceState);
        File moviesDir = Environment
                .getExternalStoragePublicDirectory(
                        Environment.DIRECTORY_MOVIES);
        outputDir = new File(moviesDir,
                "VideoRecorder");
        outputDir.mkdirs();
        setContentView(R.layout.activity_main);
    }

    @Override
    protected void onResume() {
        super.onResume();
        mediaRecorder = new MediaRecorder();
        initAndConfigureMediaRecorder();
    }
```

```java
@Override
protected void onPause() {
    super.onPause();
    if (recording) {
        try {
            mediaRecorder.stop();
        } catch (IllegalStateException e) {
        }
    }
    releaseMediaRecorder();
    Button button = (Button) findViewById(R.id.button1);
    button.setText("Start");
    recording = false;
}

private void releaseMediaRecorder() {
    if (mediaRecorder != null) {
        mediaRecorder.reset();
        mediaRecorder.release();
        mediaRecorder = null;
    }
}

private void initAndConfigureMediaRecorder() {
    mediaRecorder.setAudioSource(
            MediaRecorder.AudioSource.CAMCORDER);
    mediaRecorder
            .setVideoSource(MediaRecorder.VideoSource.CAMERA);
    mediaRecorder.setOutputFormat(
            MediaRecorder.OutputFormat.MPEG_4);
    mediaRecorder.setVideoFrameRate(10);// make it very low
    mediaRecorder.setVideoEncoder(
            MediaRecorder.VideoEncoder.MPEG_4_SP);
    mediaRecorder.setAudioEncoder(
            MediaRecorder.AudioEncoder.AMR_NB);
    String outputFile = new File(outputDir,
            System.currentTimeMillis() + ".mp4")
                .getAbsolutePath();

    mediaRecorder.setOutputFile(outputFile);
    SurfaceView surfaceView = (SurfaceView)
            findViewById(R.id.surfaceView);
    SurfaceHolder surfaceHolder = surfaceView.getHolder();
    mediaRecorder.setPreviewDisplay(surfaceHolder
            .getSurface());
}


public void startStopRecording(View view) {
    Button button = (Button) findViewById(R.id.button1);
    if (recording) {
        button.setText("Start");
```

```
        try {
            mediaRecorder.stop();
        } catch (IllegalStateException e) {

        }
        releaseMediaRecorder();
    } else {
        button.setText("Stop");
        if (mediaRecorder == null) {
            mediaRecorder = new MediaRecorder();
            initAndConfigureMediaRecorder();
        }
        // prepare MediaRecorder
        try {
            mediaRecorder.prepare();
        } catch (IllegalStateException e) {
            e.printStackTrace();
        } catch (IOException e) {
            e.printStackTrace();
        }
        mediaRecorder.start();
    }
    recording = !recording;
    }
}
```

Let's start with the **onCreate** method. It does an important job which is to create a directory for all videos captured under the default directory for movie files.

```
File moviesDir = Environment
        .getExternalStoragePublicDirectory(
            Environment.DIRECTORY_MOVIES);
outputDir = new File(moviesDir,
        "VideoRecorder");
outputDir.mkdirs();
```

The other two important methods are **onResume** and **onPause**. In **onResume** you create a new instance of **MediaRecorder** and initialize and configure it by calling **initAndConfigureMediaRecorder**. Why a new instance every time? Because once used, a **MediaRecorder** cannot be reused.

In **onPause**, you stop the **MediaRecorder** if it is recording and call the **releaseMediaRecorder** method to release the **MediaRecorder**.

Now, let's have a look at the **initAndConfigureMediaRecorder** and **releaseMediaRecorder** methods.

As the name implies, **initAndConfigureMediaRecorder** initializes and configures the **MediaRecorder** created by the **onResume** method. It calls various methods in **MediaRecorder** to transition it to the **Initialized** and **DataSourceConfigured** states. It also passes the **Surface** of the **SurfaceView** to display what the camera sees.

```
SurfaceView surfaceView = (SurfaceView)
        findViewById(R.id.surfaceView);
SurfaceHolder surfaceHolder = surfaceView.getHolder();
```

```
mediaRecorder.setPreviewDisplay(surfaceHolder
        .getSurface());
```

In this state, the **MediaRecorder** just waits until the user presses the Start button. When it happens, the **startStopRecording** method is called, which in turn calls the **prepare** and **start** methods on the **MediaRecorder**. It also changes the Start button to a Stop button.

When the user presses the Stop button, the **MediaRecorder**'s **stop** method is called and the **MediaRecorder** is released. The Stop button is changed back to a Start button, waiting for another turn.

Summary

Two methods are available if you want to equip your application with video-making capability. The first, the easy one, is by creating the default intent and passing it to **startActivityForResult**. The second method is to use **MediaRecorder** directly. This method is harder but brings with it the full features of the device camera.

This chapter showed how to use both methods to make video.

Chapter 21
The Sound Recorder

The Android platform ships with a multitude of APIs, including one for recording audio and video. In this chapter you learn how to use the **MediaRecorder** class to sample sound levels. This is the same class you used for making videos in Chapter 20, "Making Videos."

The MediaRecorder Class

Support for multimedia is rock solid in Android. There are classes that you can use to play audio and video as well as record them. In the **SoundMeter** project discussed in this chapter, you will use the **MediaRecorder** class to sample sound or noise levels.**MediaRecorder** is used to record audio and video. The output can be written to a file and the input source can be easily selected. It is relatively easy to use too. You start by instantiating the **MediaRecorder** class.

```
MediaRecorder mediaRecorder = new MediaRecorder();
```

Then, configure the instance by calling its **setAudioSource**, **setVideoSource**, **setOutputFormat**, **setAudioEncoder**, **setOutputFile**, or other methods. Next, prepare the **MediaRecorder** by calling its **prepare** method:

```
mediaRecorder.prepare();
```

Note that **prepare** may throw exception if the **MediaRecorder** is not configured property or if you do not have the right permissions.

To start recording, call its **start** method. To stop recording, call **stop**.

When you're done with a **MediaRecorder**, call its **reset** method to return it to its initial state and its **release** method to release resources it currently holds.

```
mediaRecorder.reset();
mediaRecorder.release();
```

Example

Now that you know how to use the **MediaRecorder**, let's take a look at the SoundMeter project. The application samples sound amplitudes at certain intervals and displays the current level as a bar.

As usual, let's start by looking at the manifest (the **AndroidManifest.xml** file) for the project. It is given in Listing 21.1.

Listing 21.1: The manifest for SoundMeter

```xml
<?xml version="1.0" encoding="utf-8"?>
<manifest xmlns:android="http://schemas.android.com/apk/res/android"
    package="com.example.soundmeter"
    android:versionCode="1"
    android:versionName="1.0" >

    <uses-sdk
        android:minSdkVersion="8"
        android:targetSdkVersion="17" />

    <uses-permission android:name="android.permission.RECORD_AUDIO" />

    <application
        android:allowBackup="true"
        android:icon="@drawable/ic_launcher"
        android:label="@string/app_name"
        android:theme="@style/AppTheme" >
        <activity
            android:name="com.example.soundmeter.MainActivity"
            android:label="@string/app_name" >
            <intent-filter>
                <action android:name="android.intent.action.MAIN"/>
                <category android:name="android.intent.category.LAUNCHER"
        />
            </intent-filter>
        </activity>
    </application>
</manifest>
```

One thing to note here is the use of the **uses-permission** element in the manifest to ask for the user's permission to record audio. If you don't include this element, your application will not work. Also, if the user does not consent, the application will not install.

There is only one activity in this project as can be seen in the manifest.

Listing 21.2 shows the layout file for the main activity. A **RelativeLayout** is used for the main display and it contains a **TextView** for displaying the current sound level and a button that will act as a sound indicator.

Listing 21.2: The res/layout/activity_main.xml file in SoundMeter

```xml
<RelativeLayout
        xmlns:android="http://schemas.android.com/apk/res/android"
    xmlns:tools="http://schemas.android.com/tools"
    android:layout_width="match_parent"
    android:layout_height="match_parent"
    android:paddingBottom="@dimen/activity_vertical_margin"
    android:paddingLeft="@dimen/activity_horizontal_margin"
    android:paddingRight="@dimen/activity_horizontal_margin"
    android:paddingTop="@dimen/activity_vertical_margin"
    tools:context=".MainActivity" >
```

```
<TextView
    android:id="@+id/level"
    android:layout_width="wrap_content"
    android:layout_height="wrap_content" />

<Button
    android:id="@+id/button1"
    style="?android:attr/buttonStyleSmall"
    android:layout_width="wrap_content"
    android:layout_height="wrap_content"
    android:layout_alignLeft="@+id/level"
    android:layout_below="@+id/level"
    android:background="#ff0000"
    android:layout_marginTop="30dp" />
```

```
</RelativeLayout>
```

There are two Java classes in this application. The first one, given in Listing 21.3, is a class called **SoundMeter** that encapsulates a **MediaRecorder** and exposes three methods to manage it. The first method, **start**, creates an instance of **MediaRecorder**, configures it, and starts it. The second method, **stop**, stops the **MediaRecorder**. The third method, **getAmplitude**, returns a **double** indicating the sampled sound level.

Listing 21.3: The SoundMeter class

```java
package com.example.soundmeter;
import java.io.IOException;
import android.media.MediaRecorder;

public class SoundMeter {

    private MediaRecorder mediaRecorder;
    boolean started = false;

    public void start() {
        if (started) {
            return;
        }
        if (mediaRecorder == null) {
            mediaRecorder = new MediaRecorder();

            mediaRecorder.setAudioSource(
                    MediaRecorder.AudioSource.MIC);
            mediaRecorder.setOutputFormat(
                    MediaRecorder.OutputFormat.THREE_GPP);
            mediaRecorder.setAudioEncoder(
                    MediaRecorder.AudioEncoder.AMR_NB);
            mediaRecorder.setOutputFile("/dev/null");
            try {
                mediaRecorder.prepare();
            } catch (IllegalStateException e) {
                e.printStackTrace();
            } catch (IOException e) {
```

```
                e.printStackTrace();
            }
            mediaRecorder.start();
            started = true;
        }
    }

    public void stop() {
        if (mediaRecorder != null) {
            mediaRecorder.stop();
            mediaRecorder.release();
            mediaRecorder = null;
            started = false;
        }
    }
    public double getAmplitude() {
        return mediaRecorder.getMaxAmplitude() / 100;
    }
}
```

The second Java class, **MainActivity**, is the main activity class for the application. It is presented in Listing 21.4.

Listing 21.4: The MainActivity class in SoundMeter

```
package com.example.soundmeter;
import android.app.Activity;
import android.os.Bundle;
import android.os.Handler;
import android.view.Menu;
import android.widget.Button;
import android.widget.TextView;

public class MainActivity extends Activity {
    Handler handler = new Handler();
    SoundMeter soundMeter = new SoundMeter();

    @Override
    protected void onCreate(Bundle savedInstanceState) {
        super.onCreate(savedInstanceState);
        setContentView(R.layout.activity_main);
    }

    @Override
    public boolean onCreateOptionsMenu(Menu menu) {
        // Inflate the menu; this adds items to the action bar if it
        // is present.
        getMenuInflater().inflate(R.menu.menu_main, menu);
        return true;
    }

    @Override
    public void onStart() {
        super.onStart();
```

```
        soundMeter.start();
        handler.postDelayed(pollTask, 150);
    }

    @Override
    public void onPause() {
        soundMeter.stop();
        super.onPause();
    }

    private Runnable pollTask = new Runnable() {
        @Override
        public void run() {
            double amplitude = soundMeter.getAmplitude();
            TextView textView = (TextView) findViewById(R.id.level);
            textView.setText("amp:" + amplitude);
            Button button = (Button) findViewById(R.id.button1);
            button.setWidth((int) amplitude * 10);
            handler.postDelayed(pollTask, 150);
        }
    };
}
```

The **MainActivity** class overrides two activity lifecycle methods, **onStart** and **onPause**. You may recall that the system calls **onStart** right after an activity was created or after it was restarted. The system calls **onPause** when the activity was paused because another activity was started or because an important event occurred. In the **MainActivity** class, the **onStart** method starts the **SoundMeter** and the **onPause** method stops it. The **MainActivity** class also uses a **Handler** to sample the sound level every 150 milliseconds.

Figure 21.1 shows the application. The horizontal bar shows the current sound amplitude.

Figure 21.1: The SoundMeter application

Summary

In this chapter you learned to use the **MediaRecorder** class to record audio. You also created an application for sampling noise levels.

Chapter 22
Handling the Handler

One of the most interesting and useful types in the Android SDK is the **Handler** class. Most of the time, it is used to process messages and schedule a task to run at a future time.

This chapter explains what the class is good for and offers examples.

Overview

The **android.os.Handler** class is an exciting utility class that, among others, can be scheduled do execute a **Runnable** at a future time. Any task assigned to a **Handler** will run on the **Handler**'s thread. In turn, the **Handler** runs on the thread that created it, which in most cases would be the UI thread. As such, you should not schedule a long-running task with a **Handler** because it would make your application freeze. However, you can use a **Handler** to handle a long-running task if you can be split the task into smaller parts, which you learn how to achieve in this section.

To schedule a task to run at a future time, call the **Handler** class's **postDelayed** or **postAtTime** method.

```
public final boolean postDelayed(Runnable task, long x)
```

public final boolean postAtTime(Runnable *task*, long *time*)**postDelayed** runs a task *x* milliseconds after the method is called. For example, if you want a **Runnable** to start five seconds from now, use this code.

```
Handler handler = new Handler();
```

handler.postDelayed(*runnable*, 5000);**postAtTime** runs a task at a certain time in the future. For example, if you want a task to run six seconds later, write this.

```
Handler handler = new Handler();
handler.postAtTime(runnable, 6000 + System.currentTimeMillis());
```

Example

As an example, consider the **HandlerDemo** project that uses **Handler** to animate an **ImageView**. The animation performed is simple: show an image for 400 milliseconds, then hide it for 400 milliseconds, and repeat this five times. The entire task would take about four seconds if all the work is done in a **for** loop that sleeps for 400 milliseconds at each iteration. Using the **Handler**, however, you can split this into 10 smaller parts that each takes less than one millisecond (the exact time would depend on the device running it). The UI thread is released during each 400ms wait so that it can cater for something else.

Note

Android offers animation APIs that you should use for all animation tasks. This example uses **Handler** to animate a control simply to illustrate the use of **Handler**.

Listing 22.1 shows the manifest (the **AndroidManifest.xml** file) for the project.

Listing 22.1: The manifest for HandlerDemo

```xml
<?xml version="1.0" encoding="utf-8"?>
<manifest xmlns:android="http://schemas.android.com/apk/res/android"
    package="com.example.handlerdemo"
    android:versionCode="1"
    android:versionName="1.0" >

    <uses-sdk
        android:minSdkVersion="8"
        android:targetSdkVersion="17" />

    <application
        android:allowBackup="true"
        android:icon="@drawable/ic_launcher"
        android:label="@string/app_name"
        android:theme="@style/AppTheme" >
        <activity
            android:name=".MainActivity"
            android:label="@string/app_name" >
            <intent-filter>
                <action android:name="android.intent.action.MAIN"/>
                <category
                    android:name="android.intent.category.LAUNCHER"/>
            </intent-filter>
        </activity>
    </application>
</manifest>
```

Nothing spectacular in the manifest. It shows that there is one activity named **MainActivity**. The layout file for the activity is given in Listing 22.2.

Listing 22.2: The res/layout/activity_main.xml file in HandlerTest

```xml
<RelativeLayout
    xmlns:android="http://schemas.android.com/apk/res/android"
    xmlns:tools="http://schemas.android.com/tools"
    android:layout_width="match_parent"
    android:layout_height="match_parent"
    android:paddingBottom="@dimen/activity_vertical_margin"
    android:paddingLeft="@dimen/activity_horizontal_margin"
    android:paddingRight="@dimen/activity_horizontal_margin"
    android:paddingTop="@dimen/activity_vertical_margin"
    tools:context=".MainActivity" >
```

```
<ImageView
    android:id="@+id/imageView1"
    android:layout_width="wrap_content"
    android:layout_height="wrap_content"
    android:layout_alignParentLeft="true"
    android:layout_alignParentTop="true"
    android:layout_marginLeft="51dp"
    android:layout_marginTop="58dp"
    android:src="@drawable/surprise" />

<Button
    android:id="@+id/button1"
    style="?android:attr/buttonStyleSmall"
    android:layout_width="wrap_content"
    android:layout_height="wrap_content"
    android:layout_alignRight="@+id/imageView1"
    android:layout_below="@+id/imageView1"
    android:layout_marginRight="18dp"
    android:layout_marginTop="65dp"
    android:onClick="buttonClicked"
    android:text="Button"/>
</RelativeLayout>
```

The main layout for **MainActivity** is a **RelativeLayout** that contains an **ImageView** to be animated and a button to start animation.

Now look at the **MainActivity** class in Listing 22.3. This is the main core of the application.

Listing 22.3: The MainActivity class in HandlerDemo

```
package com.example.handlerdemo;
import android.app.Activity;
import android.os.Bundle;
import android.os.Handler;
import android.view.Menu;
import android.view.View;
import android.widget.ImageView;

public class MainActivity extends Activity {

    int counter = 0;
    Handler handler = new Handler();

    @Override
    protected void onCreate(Bundle savedInstanceState) {
        super.onCreate(savedInstanceState);
        setContentView(R.layout.activity_main);
        getUserAttention();
    }

    @Override
    public boolean onCreateOptionsMenu(Menu menu) {
        // Inflate the menu; this adds items to the action bar if it
```

```
        // is present.
        getMenuInflater().inflate(R.menu.menu_main, menu);
        return true;
    }

    public void buttonClicked(View view) {
        counter = 0;
        getUserAttention();
    }

    private void getUserAttention() {
        handler.post(task);
    }

    Runnable task = new Runnable() {
        @Override
        public void run() {
            ImageView imageView = (ImageView)
                    findViewById(R.id.imageView1);
            if (counter % 2 == 0) {
                imageView.setVisibility(View.INVISIBLE);
            } else {
                imageView.setVisibility(View.VISIBLE);
            }
            counter++;
            if (counter < 8) {
                handler.postDelayed(this, 400);
            }
        }
    };
}
```

The brain of this activity are a **Runnable** called **task**, which animates the **ImageView**, and the **getUserAttention** method that calls the **postDelayed** method on a **Handler**. The **Runnable** sets the **ImageView**'s visibility to **Visible** or **Invisible** depending on whether the value of the **counter** variable is odd or even.

If you run the **HandlerDemo** project, you'll see something similar to the screenshot in Figure 22.1. Note how the **ImageView** flashes to get your attention. Try clicking the button several times quickly to make the image flash faster. Can you explain why it goes faster as you click?

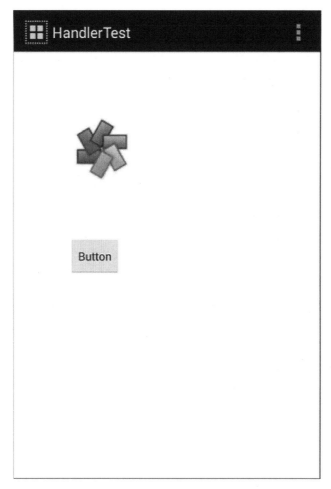

Figure 22.1: The HandlerTest application

Summary

In this chapter you learned about the **Handler** class and write an application that makes use of the class.

Chapter 23
Asynchronous Tasks

This chapter talks about asynchronous tasks and explains how to handle them using the **AsyncTask** class. It also presents a photo editor application that illustrates how this class should be used.

Overview

The **android.os.AsyncTask** class is a utility class that makes it easy to handle background processes and publish progress updates on the UI thread. This class is meant for short operations that last at most a few seconds. For long-running background tasks, you should use the Java Concurrency Utilities framework.

The **AsyncTask** class comes with a set of public methods and a set of protected methods. The public methods are for executing and canceling its task. The **execute** method starts an asynchronous operation and **cancel** cancels it. The protected methods are for you to override in a subclass. The **doInBackground** method, a protected method, is the most important method in this class and provides the logic for the asynchronous operation.

There is also a **publishProgress** method, also a protected method, which is normally called multiple times from **doInBackground**. Typically, you will write code to update a progress bar or some other UI component here.

Then there are two **onCancelled** methods for you to write what should happen if the operation was canceled (i.e. if the **AsyncTask**'s **cancel** method was called).

Example

As an example, the PhotoEditor application that accompanies this book uses the **AsyncTask** class to perform image operations that each takes a few seconds. **AsyncTask** is used so as not to jam the UI thread. Two image operations, invert and blur, are supported.

The application manifest (the **AndroidManifest.xml** file) is printed in Listing 23.1.

Listing 23.1: The manifest for PhotoEditor

```xml
<?xml version="1.0" encoding="utf-8"?>
<manifest xmlns:android="http://schemas.android.com/apk/res/android"
    package="com.example.photoeditor"
    android:versionCode="1"
```

```
    android:versionName="1.0" >

    <uses-sdk
        android:minSdkVersion="8"
        android:targetSdkVersion="17" />

    <application
        android:allowBackup="true"
        android:icon="@drawable/ic_launcher"
        android:label="@string/app_name"
        android:theme="@style/AppTheme" >
        <activity
            android:name="com.example.photoeditor.MainActivity"
            android:label="@string/app_name" >
            <intent-filter>
                <action android:name="android.intent.action.MAIN" />
                <category android:name="android.intent.category.LAUNCHER"
    />
            </intent-filter>
        </activity>
    </application>

</manifest>
```

The layout file, printed in Listing 23.2, shows that the application uses a vertical
LinearLayout to house an **ImageView**, a **ProgressBar**, and two buttons. The latter are
contained in a horizontal **LinearLayout**. The first button is used to start the blur operation
and the second to start the invert operation.

Listing 23.2: The res/layout/activity_main.xml file in PhotoEditor

```
<LinearLayout xmlns:android="http://schemas.android.com/apk/res/android"
    xmlns:tools="http://schemas.android.com/tools"
    android:layout_width="fill_parent"
    android:layout_height="fill_parent"
    android:orientation="vertical"
    android:paddingLeft="16dp"
    android:paddingRight="16dp" >

    <LinearLayout
        android:layout_height="wrap_content"
        android:layout_width="fill_parent"
        android:orientation="horizontal" >

        <Button
            android:id="@+id/blurButton"
            android:layout_width="wrap_content"
            android:layout_height="wrap_content"
            android:onClick="doBlur"
            android:text="@string/blur_button_text" />
```

```
        <Button
            android:id="@+id/button2"
            android:layout_width="wrap_content"
            android:layout_height="wrap_content"
            android:onClick="doInvert"
            android:text="@string/invert_button_text" />
    </LinearLayout>

    <ProgressBar
        android:id="@+id/progressBar1"
        style="?android:attr/progressBarStyleHorizontal"
        android:layout_width="fill_parent"
        android:layout_height="10dp" />

    <ImageView
        android:id="@+id/imageView1"
        android:layout_width="wrap_content"
        android:layout_height="wrap_content"
        android:layout_gravity="top|center"
        android:src="@drawable/photo1" />

</LinearLayout>
```

Finally, the **MainActivity** class for this project is given in Listing 23.3.

Listing 23.3: The MainActivity class in PhotoEditor

```
package com.example.photoeditor;
import android.app.Activity;
import android.graphics.Bitmap;
import android.graphics.drawable.BitmapDrawable;
import android.os.AsyncTask;
import android.os.Bundle;
import android.view.Menu;
import android.view.View;
import android.widget.ImageView;
import android.widget.ProgressBar;

public class MainActivity extends Activity {
    private ProgressBar progressBar;

    @Override
    protected void onCreate(Bundle savedInstanceState) {
        super.onCreate(savedInstanceState);
        setContentView(R.layout.activity_main);
        progressBar = (ProgressBar) findViewById(R.id.progressBar1);
    }

    @Override
    public boolean onCreateOptionsMenu(Menu menu) {
        // Inflate the menu; this adds items to the action bar if it
        // is present.
        getMenuInflater().inflate(R.menu.menu_main, menu);
        return true;
```

```java
    }

    public void doBlur(View view) {
        BlurImageTask task = new BlurImageTask();
        ImageView imageView = (ImageView)
                findViewById(R.id.imageView1);
        Bitmap bitmap = ((BitmapDrawable)
                imageView.getDrawable()).getBitmap();
        task.execute(bitmap);
    }

    public void doInvert(View view) {
        InvertImageTask task = new InvertImageTask();
        ImageView imageView = (ImageView)
                findViewById(R.id.imageView1);
        Bitmap bitmap = ((BitmapDrawable)
        imageView.getDrawable()).getBitmap();
        task.execute(bitmap);
    }

    private class InvertImageTask extends AsyncTask<Bitmap, Integer,
            Bitmap> {
        protected Bitmap doInBackground(Bitmap... bitmap) {
            Bitmap input = bitmap[0];
            Bitmap result = input.copy(input.getConfig(),
                    /*isMutable'*/true);
            int width = input.getWidth();
            int height = input.getHeight();
            for (int i = 0; i < height; i++) {
                for (int j = 0; j < width; j++) {
                    int pixel = input.getPixel(j, i);
                    int a = pixel & 0xff000000;
                    a = a | (~pixel & 0x00ffffff);
                    result.setPixel(j, i, a);
                }
                int progress = (int) (100*(i+1)/height);
                publishProgress(progress);
            }
            return result;
        }

        protected void onProgressUpdate(Integer... values) {
            progressBar.setProgress(values[0]);
        }

        protected void onPostExecute(Bitmap result) {
            ImageView imageView = (ImageView)
                    findViewById(R.id.imageView1);
            imageView.setImageBitmap(result);
            progressBar.setProgress(0);
        }
    }
```

```
private class BlurImageTask extends AsyncTask<Bitmap, Integer,
        Bitmap> {
    protected Bitmap doInBackground(Bitmap... bitmap) {
        Bitmap input = bitmap[0];
        Bitmap result = input.copy(input.getConfig(),
                /*isMutable=*/ true);
        int width = bitmap[0].getWidth();
        int height = bitmap[0].getHeight();
        int level = 7;
        for (int i = 0; i < height; i++) {
            for (int j = 0; j < width; j++) {
                int pixel = bitmap[0].getPixel(j, i);
                int a = pixel & 0xff000000;
                int r = (pixel >> 16) & 0xff;
                int g = (pixel >> 8) & 0xff;
                int b = pixel & 0xff;
                r = (r+level)/2;
                g = (g+level)/2;
                b = (b+level)/2;
                int gray = a | (r << 16) | (g << 8) | b;
                result.setPixel(j, i, gray);
            }
            int progress = (int) (100*(i+1)/height);
            publishProgress(progress);
        }
        return result;
    }

    protected void onProgressUpdate(Integer... values) {
        progressBar.setProgress(values[0]);
    }

    protected void onPostExecute(Bitmap result) {
        ImageView imageView = (ImageView)
                findViewById(R.id.imageView1);
        imageView.setImageBitmap(result);
        progressBar.setProgress(0);
    }
}
}
```

The **MainActivity** class contains two private classes, **InvertImageTask** and **BlurImageTask**, which extend **AsyncTask**. The **InvertImageTask** task is executed when the **Invert** button is clicked and the **BlurImageTask** when the **Blur** button is clicked.

The **doInBackground** method in each task processes the **ImageView** bitmap in a **for** loop. At each iteration it calls the **publishProgress** method to update the progress bar.

Figure 23.1 shows the initial bitmap and Figure 23.2 shows the bitmap after an invert operation.

Figure 23.1: The ImageEditor application

Figure 23.2: The bitmap after invert

Summary

In this chapter you learned to use the **AsyncTask** class and created a photo editor application that uses it.

Chapter 24
Services

So far in this book, everything you have learned is related to activities. It is now time to present another Android component, the service. A service has no user interface and runs in the background. It is suitable for long-running operations. This chapter explains how to create a service and provides an example.

Overview

As already mentioned, a service is a component that perform a long running operation in the background. A service will continue to run even after the application that started it has been stopped. A service runs on the same process as the application in which the service is declared and in the application's main thread. As such, if a service takes a long time to complete, it should run on a separate thread. The good thing is, running a service on a separate thread is easy if you extend a certain class in the Service API.

A service can take one of two forms. It can be started or bound. A service is started if another component starts it. It can run in the background indefinitely even after the component that started it is no longer in service or destroyed. A service is bound if an application component binds to it. A bound service acts like a server in a client-server relationship, taking requests from other application components and returning results. A service can also be started and bound.

In terms of accessibility, a service can be made private or public. A public service can be invoked by any application. A private service, on the other hand, can only be invoked by a component in the same application in which the service is declared.

The Service API

To create a service you must write a class that extends **android.app.Service** or its subclass **android.app.IntentService**. Subclassing **IntentService** is easier because it requires you to override fewer methods. However, extending **Service** allows you more control.

If you decide to subclass **Service**, you may need to override the callback methods in it. These methods are listed in Table 24.1.

Method	Description
onStartCommand	This method is called when another application component calls the service's startService to start it.
onBind	This method is invoked when another application component calles the service's bindService to bind with it.
onCreate	This method is invoked when the service is first created.
onDestroy	This method is invoked when the service is being destroyed.

Table 24.1: The Service class's callback methods

If you extend **IntentService**, you have to override its abstract method **onHandleIntent**. Here is the signature of this method.

```
protected abstract void onHandleIntent(
        android.content.Intent intent)
```

The implementation of **onHandleIntent** should contain code that needs to be executed by the service. Also note that **onHandleIntent** is always run on a separate worker thread.

Declaring A Service

A service must be declared in the manifest using the **service** element under **<application>**. The attributes that may appear in the **service** element are shown in Table 24.2.

Attribute	Description
enabled	Indicates whether the service should be enabled. The value is either true (default) or false.
exported	Accepts a value of true or false to indicate whether or not the service can be started or invoked from other applications.
icon	An icon representing this service.
isolatedProcess	Accepts a value of true or false indicating whether the service should be run as a separate process.
label	A label for this service.
name	The fully-qualified name of the service class.
permission	The name of a permission that that an entity must have in order to launch the service or bind to it.
process	The name of the process where the service is to run.

Table 24.2: The attributes of the service element

For example, here is the declaration of a **service** element that can be invoked by other applications.

```
<application>
    ...

    <service android:name="com.example.MyService"
        android:exported="true" />
```

```
</application>
```

A Service Example

This example is an Android application that lets you download web pages and store them for offline viewing when you have no Internet access.

There are two activities and one service. In the main activity (shown in Figure 24.1), you can enter URLs of the web sites whose contents you want to store in your device. Just type a URL in each line and click the **FETCH WEB PAGES** button to start the URL Service.

Figure 24.1: The Main activity

Click VIEW PAGES on the action bar to view the stored contents. You will see the second activity like that shown in Figure 24.2.

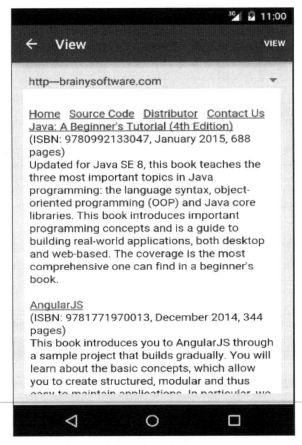

Figure 24.2: The View activity

The View activity's view area consists of a Spinner and a WebView. The spinner contains encoded URLs that have been fetched. Select a URL to display the content in the WebView.

Now that you have an idea of what the app does, let's take a look at the code.

As usual, you start from the manifest. It describes the application and is listed in Listing 24.1.

Listing 24.1: The manifest
```xml
<?xml version="1.0" encoding="utf-8"?>
<manifest xmlns:android="http://schemas.android.com/apk/res/android"
    package="com.example.urlservice" >

    <uses-permission android:name="android.permission.INTERNET" />
    <uses-permission
    android:name="android.permission.ACCESS_NETWORK_STATE" />

    <application
        android:allowBackup="true"
        android:icon="@drawable/ic_launcher"
```

```
            android:label="@string/app_name"
            android:theme="@style/AppTheme" >
            <activity
                android:name=".MainActivity"
                android:label="@string/app_name" >
                <intent-filter>
                    <action android:name="android.intent.action.MAIN" />
                    <category
➡android:name="android.intent.category.LAUNCHER" />
                </intent-filter>
            </activity>

            <activity
                android:name=".ViewActivity"
                android:parentActivityName=".MainActivity"
                android:label="@string/title_activity_view" >
            </activity>

            <service
                android:name=".URLService"
                android:exported="true" />
        </application>
</manifest>
```

As you can see the **application** element contains two **activity** elements and a **service** element. There are also two **uses-permission** elements to give the application to access the Internet. They are **android.permission. INTERNET** and **android.permission.ACCESS_NETWORK_STATE**.

Listing 24.1: The main activity class

```
package com.example.urlservice;
import android.content.Intent;
import android.os.StrictMode;
import android.support.v7.app.ActionBarActivity;
import android.os.Bundle;
import android.util.Log;
import android.view.Menu;
import android.view.MenuItem;
import android.view.View;
import android.widget.EditText;

public class MainActivity extends ActionBarActivity {

    @Override
    protected void onCreate(Bundle savedInstanceState) {
        super.onCreate(savedInstanceState);
        setContentView(R.layout.activity_main);
        StrictMode.ThreadPolicy policy = new
                StrictMode.ThreadPolicy.Builder().permitAll().build();
        StrictMode.setThreadPolicy(policy);
    }
```

```java
    @Override
    public boolean onCreateOptionsMenu(Menu menu) {
        getMenuInflater().inflate(R.menu.menu_main, menu);
        return true;
    }


    @Override
    public boolean onOptionsItemSelected(MenuItem item) {
        int id = item.getItemId();
        if (id == R.id.action_view) {
            Intent intent = new Intent(this, ViewActivity.class);
            startActivity(intent);
            return true;
        }
        return super.onOptionsItemSelected(item);
    }

    public void fetchWebPages(View view) {
        EditText editText = (EditText) findViewById(R.id.urlsEditText);
        Intent intent = new Intent(this, URLService.class);
        intent.putExtra("urls", editText.getText().toString());
        startService(intent);
    }
}
```

Listing 24.2: The view activity class

```java
package com.example.urlservice;
import android.os.Bundle;
import android.support.v7.app.ActionBarActivity;
import android.view.Menu;
import android.view.MenuItem;
import android.view.View;
import android.webkit.WebView;
import android.widget.AdapterView;
import android.widget.ArrayAdapter;
import android.widget.Spinner;
import java.io.BufferedReader;
import java.io.File;
import java.io.FileNotFoundException;
import java.io.FileReader;
import java.io.IOException;


public class ViewActivity extends ActionBarActivity {

    @Override
    protected void onCreate(Bundle savedInstanceState) {
        super.onCreate(savedInstanceState);
        setContentView(R.layout.activity_view);
        Spinner spinner = (Spinner) findViewById(R.id.spinner);
        File saveDir = getFilesDir();
```

```java
        if (saveDir.exists()) {
            File dir = new File(saveDir, "URLService");
            dir = saveDir;
            if (dir.exists()) {
                String[] files = dir.list();
                ArrayAdapter<String> dataAdapter =
                        new ArrayAdapter<String>(this,
                        android.R.layout.simple_spinner_item, files);
                dataAdapter.setDropDownViewResource(
                        android.R.layout.simple_spinner_dropdown_item);
                spinner.setAdapter(dataAdapter);
                spinner.setOnItemSelectedListener(
                        new AdapterView.OnItemSelectedListener() {
                    @Override
                    public void onItemSelected(AdapterView<?>
                                adapterView, View view, int pos,
                                long id) {
                        //open file
                        Object itemAtPosition = adapterView
                                .getItemAtPosition(pos);
                        File file = new File(getFilesDir(),
                                itemAtPosition.toString());
                        FileReader fileReader = null;
                        BufferedReader bufferedReader = null;
                        try {
                            fileReader = new FileReader(file);
                            bufferedReader =
                                    new BufferedReader(fileReader);
                            StringBuilder sb = new StringBuilder();
                            String line = bufferedReader.readLine();
                            while (line != null) {
                                sb.append(line);
                                line = bufferedReader.readLine();
                            }
                            WebView webView = (WebView)
                                    findViewById(R.id.webview);
                            webView.loadData(sb.toString(),
                                    "text/html", "utf-8");
                        } catch (FileNotFoundException e) {
                        } catch (IOException e) {
                        }
                    }

                    @Override
                    public void onNothingSelected(AdapterView<?>
                            adapterView) {
                    }
                });
            }
        }
    }

    @Override
```

```
    public boolean onCreateOptionsMenu(Menu menu) {
        getMenuInflater().inflate(R.menu.menu_view, menu);
        return true;
    }

    @Override
    public boolean onOptionsItemSelected(MenuItem item) {
        int id = item.getItemId();
        return super.onOptionsItemSelected(item);
    }
}
```

The most important piece of the application, the service class, is shown in Listing 24.3. It extends **IntentService** and implements its **onHandleIntent** method.

Listing 24.3: The service class

```
package com.example.urlservice;
import android.app.IntentService;
import android.content.Intent;
import java.io.BufferedReader;
import java.io.File;
import java.io.InputStreamReader;
import java.io.PrintWriter;
import java.net.MalformedURLException;
import java.net.URL;
import java.util.StringTokenizer;

public class URLService extends IntentService {
    public URLService() {
        super("URLService");
    }

    @Override
    protected void onHandleIntent(Intent intent) {
        String urls = intent.getStringExtra("urls");
        if (urls == null) {
            return;
        }
        StringTokenizer tokenizer = new StringTokenizer(urls);
        int tokenCount = tokenizer.countTokens();
        int index = 0;
        String[] targets = new String[tokenCount];
        while (tokenizer.hasMoreTokens()) {
            targets[index++] = tokenizer.nextToken();
        }
        File saveDir = getFilesDir();
        fetchPagesAndSave(saveDir, targets);
    }

    private void fetchPagesAndSave(File saveDir, String[] targets) {
        for (String target : targets) {
            URL url = null;
            try {
```

```
                url = new URL(target);
            } catch (MalformedURLException e) {
                e.printStackTrace();
            }
            String fileName = target.replaceAll("/", "-")
                    .replaceAll(":", "-");

            File file = new File(saveDir, fileName);
            PrintWriter writer = null;
            BufferedReader reader = null;
            try {
                writer = new PrintWriter(file);
                reader = new BufferedReader(
                        new InputStreamReader(url.openStream()));
                String line;
                while ((line = reader.readLine()) != null) {
                    writer.write(line);
                }
            } catch (Exception e) {
            } finally {
                if (writer != null) {
                    try {
                        writer.close();
                    } catch (Exception e) {
                    }
                }
                if (reader != null) {
                    try {
                        reader.close();
                    } catch (Exception e) {
                    }
                }
            }
        }
    }
}
```

The onHandleIntent method receives an array of URLs and uses a StringTokenizer to extract each URL from the array. Each URL is used to populate a string array named **targets**, which is then passed to the fetchPagesAndSave method. This method employs a **java.net.URL** to send an HTTP request for each target and saves its content in internal storage.

Summary

A service is an application component that runs in the background. Despite the fact that it runs in the background, a service is not a process and does not run on a separate thread. Instead, a service runs on the main thread of the application that invoked the service.

You can write a service by extending **android.app.Service** or **android.app.IntentService**.

Chapter 25
Broadcast Receivers

The Android system constantly broadcasts intents that occur during the running of the operating system and applications. In addition, applications can also broadcast user-defined intents. You can capitalize on these broadcasts by writing broadcast receivers in your application.

This chapter explains how to create broadcast receivers.

Overview

A broadcast receiver, or a receiver for short, is an application component that listens to a certain intent broadcast, similar to Java listeners that listen to events. Table 25.1 shows intent actions defined in the **android.content.Intent** class for which you can write a receiver.

Action	Description
ACTION_TIME_TICK	The current time has changed. Sent every minute.
ACTION_TIME_CHANGED	The time has been set.
ACTION_TIMEZONE_CHANGED	The timezone has changed.
ACTION_BOOT_COMPLETED	The system has finished booting.
ACTION_PACKAGE_ADDED	A new application package has been installed on the device.
ACTION_PACKAGE_CHANGED	An application package has been changed.
ACTION_PACKAGE_REMOVED	An application package has been removed.
ACTION_PACKAGE_RESTARTED	The user has restarted a package.
ACTION_PACKAGE_DATA_CLEARED	The user has cleared the data of a package.
ACTION_UID_REMOVED	A user UID has been removed.
ACTION_BATTERY_CHANGED	The battery's charging state, level or other detail has changed.
ACTION_POWER_CONNECTED	External power has been connected to the device.
ACTION_POWER_DISCONNECTED	External power has been disconnected from the device.
ACTION_SHUTDOWN	The device is about to shut down

Table 25.1: Intent actions for receiving a broadcast

To create a receiver, you must extend the **android.content.BroadcastReceiver** class or one of its subclasses. In your class, you must provide an implementation for the **onReceive** method, which gets called when an intent for which the receiver is registered is broadcast. The signature of **onReceive** is as follows.

```
public abstract void onReceive (Context context, Intent intent)
```

You then have to register your class in the application manifest using the **receiver** element or programmatically by calling **Context.registerReceiver()**.

BroadcastReceiver-based Clock

Android comes with widgets that can show time. However, you can also create your own clock widget that is based on the ACTION_TIME_TICK broadcast. Recall that this intent action is broadcast every minute, which is suitable for a clock.

The BroadcastReceiverDemo1 project features such a clock. It is a simple app that consists of a broadcast receiver and an activity. The receiver class is instantiated and registered every time the activity's **onResume** method is called. It is deregistered when **onPause** is invoked.

The class for the main activity is given in Listing 25.1

Listing 25.1: The MainActivity class

```
package com.example.broadcastreceiverdemo1;
import java.util.Calendar;
import android.app.Activity;
import android.content.BroadcastReceiver;
import android.content.Context;
import android.content.Intent;
import android.content.IntentFilter;
import android.os.Bundle;
import android.text.format.DateFormat;
import android.util.Log;
import android.view.Menu;
import android.widget.TextView;

public class MainActivity extends Activity {

    BroadcastReceiver receiver;

    @Override
    protected void onCreate(Bundle savedInstanceState) {
        super.onCreate(savedInstanceState);
        setContentView(R.layout.activity_main);
    }

    @Override
    public void onResume() {
        super.onResume();
        setTime();
        receiver = new BroadcastReceiver() {
            @Override
            public void onReceive(Context context, Intent intent) {
                setTime();
            }
```

```
        };
        IntentFilter intentFilter = new IntentFilter(
                Intent.ACTION_TIME_TICK);
        this.registerReceiver(receiver, intentFilter);
    }

    public void onPause() {
        this.unregisterReceiver(receiver);
        super.onPause();
    }
    @Override
    public boolean onCreateOptionsMenu(Menu menu) {
        getMenuInflater().inflate(R.menu.menu_main, menu);
        return true;
    }

    private void setTime() {
        Calendar calendar = Calendar.getInstance();
        CharSequence newTime = DateFormat.format(
                "kk:mm", calendar);
        TextView textView = (TextView) findViewById(
                R.id.textView1);
        textView.setText(newTime);
    }
}
```

An important part of the application is the **onReceive** method of the receiver:

```
@Override
public void onReceive(Context context, Intent intent) {
    setTime();
}
```

It is very simple method with one line of code that calls the **setTime** method. The **setTime** method obtains the current time from a **Calendar** and updates a **TextView**.

Another important part of the application is the code that registers the receiver in the activity's **onResume** method. To register a receiver, you need to create an **IntentFilter** specifying an intent action that will cause the receiver to be triggered. In this case the intent action is ACTION_TIME_TICK.

```
IntentFilter intentFilter = new IntentFilter(
        Intent.ACTION_TIME_TICK);
this.registerReceiver(receiver, intentFilter);
```

You then pass the receiver and the **IntentFilter** to register the receiver.

Figure 25.1 shows the broadcast receiver-based clock.

Figure 25.1: A receiver-based clock

Canceling A Notification

Chapter 3, "UI Components" explains the various Android UI components including notifications. A problem lingers: Touching a notification's action UI does not cancel the notification. One strategy to solve this issue is by sending a user-defined broadcast when the action UI is touched and writing a broadcast for that.

Recall that a notification action requires a **PendingIntent** and a **PendingIntent** can be programmed to send a broadcast. To solve the problem, create a user-defined intent action called **cancel_notification** and the corresponding **PendingIntent**:

```
Intent cancelIntent = new Intent("cancel_notification");
PendingIntent cancelPendingIntent =
        PendingIntent.getBroadcast(this, 100, cancelIntent, 0);
```

This **PendingIntent** can then be used to register a notification.

The CancelNotificationDemo project shows how this can be achieved. The application is made simple and consists of an activity that contains a broadcast receiver.

The layout file for the main activity is given in Listing 25.2.

Listing 25.2: The layout file for the main activity

```
<LinearLayout
        xmlns:android="http://schemas.android.com/apk/res/android"
        xmlns:tools="http://schemas.android.com/tools"
    android:layout_width="wrap_content"
    android:layout_height="wrap_content"
    android:orientation="horizontal">

    <Button
        android:layout_width="wrap_content"
        android:layout_height="wrap_content"
        android:onClick="setNotification"
        android:text="Set Notification" />

    <Button
        android:layout_width="wrap_content"
        android:layout_height="wrap_content"
        android:onClick="clearNotification"
        android:text="Clear Notification" />
</LinearLayout>
```

The layout features two buttons, one for setting a notification and one for cancaling it.

The **MainActivity** class for the application is listed in Listing 25.3. The activity's **onCreate** method instantiates a receiver whose **onReceive** method cancels the notification.

Listing 25.3: The MainActivity class

```
package com.example.cancelnotificationdemo;
import android.app.Activity;
import android.app.Notification;
import android.app.NotificationManager;
import android.app.PendingIntent;
import android.content.BroadcastReceiver;
import android.content.Context;
import android.content.Intent;
import android.content.IntentFilter;
import android.os.Bundle;
import android.util.Log;
import android.view.Menu;
import android.view.MenuItem;
import android.view.View;

public class MainActivity extends Activity {
    private static final String CANCEL_NOTIFICATION_ACTION
            = "cancel_notification";
    int notificationId = 1002;

    @Override
    protected void onCreate(Bundle savedInstanceState) {
```

```java
        super.onCreate(savedInstanceState);
        setContentView(R.layout.activity_main);

        BroadcastReceiver receiver = new BroadcastReceiver() {
            @Override
            public void onReceive(Context context, Intent intent) {
                NotificationManager notificationManager =
                        (NotificationManager) getSystemService(
                                NOTIFICATION_SERVICE);
                notificationManager.cancel(notificationId);
            }
        };
        IntentFilter filter = new IntentFilter();
        filter.addAction(CANCEL_NOTIFICATION_ACTION);
        this.registerReceiver(receiver, filter);
    }

    @Override
    public boolean onCreateOptionsMenu(Menu menu) {
        getMenuInflater().inflate(R.menu.menu_main, menu);
        return true;
    }


    public void setNotification(View view) {
        Intent cancelIntent = new Intent("cancel_notification");
        PendingIntent cancelPendingIntent =
                PendingIntent.getBroadcast(this, 100,
                        cancelIntent, 0);

        Notification notification  = new Notification.Builder(this)
                .setContentTitle("Stop Press")
                .setContentText(
                        "Everyone gets extra vacation week!")
                .setSmallIcon(android.R.drawable.star_on)
                .setAutoCancel(true)
                .addAction(android.R.drawable.btn_dialog,
                        "Dismiss", cancelPendingIntent)
                .build();

        NotificationManager notificationManager =
                (NotificationManager) getSystemService(
                        NOTIFICATION_SERVICE);
        notificationManager.notify(notificationId, notification);
    }

    public void clearNotification(View view) {
        NotificationManager notificationManager =
                (NotificationManager) getSystemService(
                        NOTIFICATION_SERVICE);
        notificationManager.cancel(notificationId);
    }
}
```

Again, note the part that register the receiver:

```
IntentFilter filter = new IntentFilter();
filter.addAction(CANCEL_NOTIFICATION_ACTION);
this.registerReceiver(receiver, filter);
```

Here, I create an **IntentFilter** that specifies a user-defined action (cancel_notification) and pass it along with the receiver to the **registerReceiver** method.

The main activity is shown in Figure 25.2.

Figure 25.2: CancelNotificationDemo

Now touch on the Set Notification button and open the notification drawer. You should see a notification like that shown in Figure 25.3.

Figure 25.3: The notification drawer

If you touch on the Dismiss button, a broadcast will be sent and received by the receiver in the activity. As a result, the notification will be canceled.

Summary

A broadcast receiver is an application component that listens to intent broadcasts. To create a receiver you must create a class that extends android.content.BroadcastReceiver and implements its onReceive method. To register a receiver, you can either add a **receiver** element in the application manifest or do so programmatically by calling **Context.registerReceiver()**. In either case, you must define an **IntentFilter** that specifies what intent should cause the receiver to be triggered.

Chapter 26
The Alarm Service

Android devices maintain an internal alarm service that can be used to schedule jobs. Amazingly, as you will find out in this chapter, the API is very easy to use, seamlessly hiding the complexity of its lower-level code. This chapter explains how to use it and presents an example.

Overview

One of the built-in services available to all Android developers is the alarm service. With it you can schedule an action to take place at a later time. The operation can be programmed to be carried out once or repeatedly. The Clock application, for example, includes an alarm clock that relies on this service.

It is extremely easy to use. All you need is encapsulate the operation you intend to schedule in a **PendingIntent** and pass it to the system-wide **AlarmManager** instance. The **AlarmManager** class is part of the **android.app** package and an instance is already there, maintained by the system. You can retrieve the **AlarmManager** by using this line of code:

```
AlarmManager alarmMgr =
        (AlarmManager) getSystemService(Context.ALARM_SERVICE);
```

The **PendingIntent** is explained in Chapter 3, "UI Components," but basically it is an intent to be invoked at a future time, hence the name **PendingIntent**. You can use a **PendingIntent** to start an activity, start a service, or broadcast a notification.

To schedule a job, call the **set** or **setExact** method of **AlarmManager**. Their signatures are as follows.

```
public void set(int type, long triggerTime, PendingIntent operation)
```

```
public void setExact(int type, long triggerTime,
        PendingIntent operation)
```

As the name implies, **setExact** causes the system to try to deliver the alarm as close as possible to the specified trigger time. On the other hand, the delivery of the job passed to **set** may be deferred but will not be earlier.

For both methods, the type is one of the following constants declared in **AlarmManager**.

- **ELAPSED_REALTIME**. The trigger time is a long representing the number of milliseconds that have elapsed since the last boot. It does no wake up the device if the alarm goes off while the device is asleep.
- **ELAPSED_REALTIME_WAKEUP**. The trigger time is a long representing the number of milliseconds that have elapsed since the last boot It wakes up the device if the alarm goes off while the device is asleep.
- **RTC**. The trigger time is a long representing the number of milliseconds that have elapsed since January 1, 1970 00:00:00.0 UTC. It does not wake up the device if the alarm goes off while the device is asleep.
- **RTC_WAKEUP**. The trigger time is a long representing the number of milliseconds that have elapsed since January 1, 1970 00:00:00.0 UTC. It wakes up the device if the alarm goes off while the device is asleep.

For example, to schedule an job to start five seconds from now, use this:

```
alarmManager.set(AlarmManager.RTC, System.currentTimeMillis() +
        5000, pendingIntent)
```

To schedule a repeating job, use the **setRepeating** or **setInexactRepeating** method. The signatures of these methods are as follows.

```
public void setInexactRepeating(int type, long triggerAtMillis,
        long intervalMillis, PendingIntent operation)

public void setRepeating(int type, long triggerAtMillis,
        long intervalMillis, PendingIntent operation)
```

In Android API levels lower than 19, **setRepeating** delivers an exact delivery time. However, starting the API level 19, **setRepeating** is also inexact, so it is the same as **setInexactRepeating**. For an exact repeating job, schedule it with **setExact**, and schedule a new job at the end of the execution of the current job.

Example

The following example shows how to schedule an alarm that sets off in five minutes. This is like the alarm clock in the Clock application, but setting an alarm is as simple as a touch of a button. The application also shows how to wake up an activity when the alarm sets off while the device is asleep.

The application has two activities, which are declared in the manifest in Listing 26.1. The first activity is the main activity that that will be launched when the user touches on the application icon on the Home screen. The second activity, called **WakeUpActivity**, is the activity that will be started when an alarm sets off.

Listing 26.1: The manifest

```
<?xml version="1.0" encoding="utf-8"?>
<manifest
    xmlns:android="http://schemas.android.com/apk/res/android"
    package="com.example.alarmmanagerdemo1" >

    <uses-permission android:name="android.permission.WAKE_LOCK"/>
```

```
    <application
        android:allowBackup="true"
        android:icon="@drawable/ic_launcher"
        android:label="@string/app_name"
        android:theme="@style/AppTheme" >
        <activity
            android:name=".MainActivity"
            android:label="@string/app_name" >
            <intent-filter>
                <action android:name="android.intent.action.MAIN" />
                <category
android:name="android.intent.category.LAUNCHER"/>
            </intent-filter>
        </activity>

        <activity
            android:name=".WakeUpActivity"
            android:label="@string/title_activity_wake_up" >
        </activity>

    </application>
</manifest>
```

The main activity contains a button, which the user can press to set an alarm. The activity layout file is shown in Listing 26.2. Note that the button declaration includes the onClick attribute that refers the a **setAlarm** method.

Listing 26.2: The layout file of the main activity

```
<RelativeLayout
    xmlns:android="http://schemas.android.com/apk/res/android"
    xmlns:tools="http://schemas.android.com/tools"
    android:layout_width="match_parent"
    android:layout_height="match_parent"
    android:paddingLeft="@dimen/activity_horizontal_margin"
    android:paddingRight="@dimen/activity_horizontal_margin"
    android:paddingTop="@dimen/activity_vertical_margin"
    android:paddingBottom="@dimen/activity_vertical_margin"
    tools:context=".MainActivity">

    <Button
        android:layout_width="wrap_content"
        android:layout_height="wrap_content"
        android:text="5 Minute Alarm"
        android:id="@+id/button"
        android:layout_alignParentLeft="true"
        android:layout_alignParentStart="true"
        android:layout_marginTop="77dp"
        android:onClick="setAlarm"/>

</RelativeLayout>
```

Listing 26.3 presents the **MainActivity** class for the application.

Listing 26.3: The MainActivity class

```java
package com.example.alarmmanagerdemo1;
import android.app.Activity;
import android.app.AlarmManager;
import android.app.PendingIntent;
import android.content.Context;
import android.content.Intent;
import android.os.Bundle;
import android.view.Menu;
import android.view.MenuItem;
import android.view.View;
import android.widget.Toast;
import java.util.Calendar;
import java.util.Date;

public class MainActivity extends Activity {

    @Override
    protected void onCreate(Bundle savedInstanceState) {
        super.onCreate(savedInstanceState);
        setContentView(R.layout.activity_main);
    }

    @Override
    public boolean onCreateOptionsMenu(Menu menu) {
        getMenuInflater().inflate(R.menu.menu_main, menu);
        return true;
    }

    @Override
    public boolean onOptionsItemSelected(MenuItem item) {
        int id = item.getItemId();
        if (id == R.id.action_settings) {
            return true;
        }
        return super.onOptionsItemSelected(item);
    }

    public void setAlarm(View view) {
        Calendar calendar = Calendar.getInstance();
        calendar.add(Calendar.MINUTE, 5);
        Date fiveMinutesLater = calendar.getTime();
        Toast.makeText(this, "The alarm will set off at " +
                fiveMinutesLater, Toast.LENGTH_LONG).show();
        Intent intent = new Intent(this, WakeUpActivity.class);
        PendingIntent sender = PendingIntent.getActivity(
                this, 0, intent, 0);
        AlarmManager alarmMgr = (AlarmManager) getSystemService(
                Context.ALARM_SERVICE);
        alarmMgr.set(AlarmManager.RTC_WAKEUP,
                fiveMinutesLater.getTime(), sender);
    }
```

```
}
```

Look at the **setAlarm** method in the **MainActivity** class. After creating a **Date** that points
to a time five minutes from now, the method creates a **PendingIntent** encapsulating an
Intent that will launch the **WakeUpActivity** activity.

```
Intent intent = new Intent(this, WakeUpActivity.class);
PendingIntent sender = PendingIntent.getActivity(
        this, 0, intent, 0);
```

It then retrieves the **AlarmManager** and set an alarm by passing the time and the
PendingIntent.

```
AlarmManager alarmMgr = (AlarmManager) getSystemService(
        Context.ALARM_SERVICE);
alarmMgr.set(AlarmManager.RTC_WAKEUP,
        fiveMinutesLater.getTime(), sender);
```

Finally, Listing 26.4 shows the **WakeUpActivity** class.

Listing 26.4: The WakeUpActivity class

```
package com.example.alarmmanagerdemo1;

import android.app.Activity;
import android.app.Notification;
import android.app.NotificationManager;
import android.os.Bundle;
import android.util.Log;
import android.view.Menu;
import android.view.MenuItem;
import android.view.View;
import android.view.Window;
import android.view.WindowManager;


public class WakeUpActivity extends Activity {
    private final int NOTIFICATION_ID = 1004;

    @Override
    protected void onCreate(Bundle savedInstanceState) {
        super.onCreate(savedInstanceState);
        final Window window = getWindow();
        Log.d("wakeup", "called. oncreate");
        window.addFlags(
                WindowManager.LayoutParams.FLAG_SHOW_WHEN_LOCKED
                | WindowManager.LayoutParams.FLAG_DISMISS_KEYGUARD
                | WindowManager.LayoutParams.FLAG_TURN_SCREEN_ON);
        setContentView(R.layout.activity_wake_up);
        addNotification();
    }

    @Override
    public boolean onCreateOptionsMenu(Menu menu) {
        getMenuInflater().inflate(R.menu.menu_wake_up, menu);
```

```
        return true;
    }

    @Override
    public boolean onOptionsItemSelected(MenuItem item) {
        int id = item.getItemId();
        if (id == R.id.action_settings) {
            return true;
        }
        return super.onOptionsItemSelected(item);
    }

    public void dismiss(View view) {
        NotificationManager notificationMgr = (NotificationManager)
                getSystemService(NOTIFICATION_SERVICE);
        notificationMgr.cancel(NOTIFICATION_ID);
        this.finish();
    }

    private void addNotification() {
        NotificationManager notificationMgr = (NotificationManager)
                getSystemService(NOTIFICATION_SERVICE);
        Notification notification  = new Notification.Builder(this)
                .setContentTitle("Wake up")
                .setSmallIcon(android.R.drawable.star_on)
                .setAutoCancel(false)
                .build();
        notification.defaults|= Notification.DEFAULT_SOUND;
        notification.defaults|= Notification.DEFAULT_LIGHTS;
        notification.defaults|= Notification.DEFAULT_VIBRATE;
        notification.flags |= Notification.FLAG_INSISTENT;
        notification.flags |= Notification.FLAG_AUTO_CANCEL;
        notificationMgr.notify(NOTIFICATION_ID, notification);
    }
}
```

At first blush, the **WakeUpActivity** class looks like other activities you've written so far, but take a close look at closely at the **onCreate** method. The following code that adds flags to the window is needed to wake up the device and show the activity if the device is asleep when the alarm sets off.

```
final Window window = getWindow();
Log.d("wakeup", "called. oncreate");
window.addFlags(
        WindowManager.LayoutParams.FLAG_SHOW_WHEN_LOCKED
        | WindowManager.LayoutParams.FLAG_DISMISS_KEYGUARD
        | WindowManager.LayoutParams.FLAG_TURN_SCREEN_ON);
setContentView(R.layout.activity_wake_up);
```

It then calls the private **addNotification** method to add a notification. Remember that Chapter 3, "UI Components" explains how to use notifications.

Figure 26.1 shows the application's main activity. Touch the button to set an alarm.

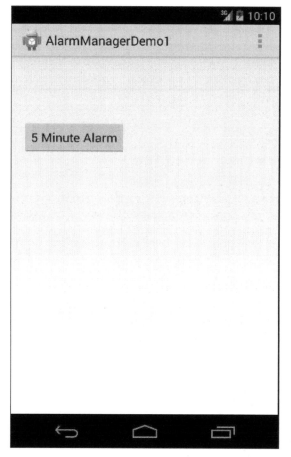

Figure 26.1: A 5-minute alarm

Summary

The alarm service is one of the built-in services available to Android developers. With it you can schedule an action to take place at a later time. The operation can be programmed to be carried out once or repeatedly.

Chapter 27
Content Providers

A content provider is an Android component used for encapsulating data that is to be shared with other applications. How the actual data is stored, be it in a relational database or a file or a mix of both, is not important. What is important is that a content provider offers a standard way of accessing data in other applications.

This chapter discusses the content provider and explains how to access data in a provider using a content resolver.

Overview

You already learned how to store files and data in a relational database. If your data needs to be shared with other applications, you need a content provider that encapsulates the stored data. Do not use a content provider if your data is to be consumed only by other components in the same application.

To create a content provider, you extend the **android.content.ContentProvider** class. This class offers CRUD methods, namely methods for creating, retrieving, updating and deleting data. Then, the subclass of **ContentProvider** has to be registered in the application manifest using the **provider** element, located under **<application>**. This class is discussed further in the next section.

Once a content provider is registered, components under the same application can access it, but not other applications. To offer the data to other applications, you must declare a read permission and a write permission. Alternatively, you can declare one permission for both read and write. Here is an example:

```
<provider
    android:name=".provider.ElectricCarContentProvider"
    android:authorities="com.example.contentproviderdemo1"
    android:enabled="true"
    android:exported="true"
    android:readPermission="com.example.permission.READ_DATA"
    android:writePermission="com.example.permission.WRITE_DATA">
</provider>
```

In addition, you need to use the permission element to re-declare the permissions in your manifest:

```
<permission
    android:name="com.example.permission.READ_DATA"
    android:protectionLevel="normal"/>
```

```
<permission
    android:name="com.example.permission.WRITE_DATA"
    android:protectionLevel="normal"/>

<application ... />
```

The permission names can be anything as long as it does not conflict with the existing ones. As such, it is a good idea to include your domain as part of your permission names.

Data in a content provider is referenced by a unique URI. The consumers of a content provider must know this URI in order to access the content provider's data.

The application containing a content provider does not need to be running for its data to be accessed.

Android comes with a number of default content providers, such as Calendar, Contacts, WordDictionary, etc. To access a content provider, you use the **android.content.ContentResolver** object that you can retrieve by calling **Context.getContentResolver()**. Among methods in the **ContentResolver** class are methods with identical names as the CRUD methods in the **ContentProvider** class. Calling one of these methods on the **ContentResolver** invokes the identically-named method in the target **ContentProvider**.

An application needing access to data in a content provider must declare that it intends to use the data, so the user installing the app is aware of what data will be exposed to the application. The consuming application must use the **uses-permission** element in its manifest. Here is an example.

```
<uses-permission
    android:name="com.example.permission.READ_DATA"/>
<uses-permission
    android:name="com.example.permission.WRITE_DATA"/>
```

The ContentProvider Class

This section introduces the CRUD methods in the **ContentProvider** class. Primarily, you need to know how to access the underlying data when overriding these methods. You can store the data in any format, but, as you will soon find out, it makes perfect sense to store the data in a relational database.

The data in a content provider is identified by URIs having this format:

```
content://authority/table
```

The authority serves as an Android internal name and should be your domain name is reverse. Right after it is the table name.

To refer to a single data item, you use this format:

```
content://authority/table/index
```

For example, suppose the authority is **com.example.provider** and the data is stored in a relational database table named customers, the first row is identified by this URI:

```
content://com.example.provider/customers/1
```

The rest of the section discusses **ContentProvider** methods for accessing and manipulating the underlying data.

The query Method

To access the underlying data, use the **query** method. Here is its signature:

```
public abstract android.database.Cursor query (android.net.Uri uri,
        java.lang.String[] projection, java.lang.String selection,
        java.lang.String[] selectionArgs,
        java.lang.String sortOrder)
```

uri is the URI identifying the data. The projection is an array containing names of the columns to be included. The selection defines which data items to select and the selectionArgs contains arguments for the selection. Finally, the sortOrder defines the column based on which the data is to be sorted.

The insert Method

The **insert** method is called to add a data item. The signature of this method is as follows.

```
public abstract android.net.Uri insert(android.net.Uri uri,
        ContentValues values)
```

You pass column key/value pairs in a **ContentValues** object to this method. Use the **put** methods of **ContentValues** to add a key/value pair.

The update Method

You use this method to update a data item or a set of data items. The signature of the method allows you to pass new values in a ContentValues as well as a selection to determine which data items will be affected. Here is the signature of **update**.

```
public abstract int update(android.net.Uri uri,
        ContentValues values, java.lang.String selection,
        java.lang.String[] selectionArgs)
```

The **update** method returns the number of data items affected.

The delete Method

The delete method deletes a data item or a set of data items. You can pass a selection and selection arguments to tell the content provider which data items should be deleted. Here is the signature of **delete**.

```
public abstract int delete(android.net.Uri uri,
        java.lang.String selection,
        java.lang.String[] selectionArgs)
```

The **delete** method returns the number of records deleted.

Creating A Content Provider

The ContentProviderDemo1 project is an application that contains a provider and three activities. The app is for green car enthusiasts and allows the user to manage electric cars. The underlying data is stored in a SQLite database. As the activities are in the same application as the provider, they do not need special permissions to access the data. The ContentResolverDemo1 project in the next section demonstrates how to access the content provider from a different application.

As always, I will start by showing the application manifest, which is given in Listing 27.1.

Listing 27.1: The manifest of ContentProviderDemo1

```xml
<?xml version="1.0" encoding="utf-8"?>
<manifest xmlns:android="http://schemas.android.com/apk/res/android"
    package="com.example.contentproviderdemo1" >

    <permission
        android:name="com.example.permission.READ_ELECTRIC_CARS"
        android:protectionLevel="normal"/>
    <permission
        android:name="com.example.permission.WRITE_ELECTRIC_CARS"
        android:protectionLevel="normal"/>

    <application
        android:allowBackup="true"
        android:icon="@drawable/ic_launcher"
        android:label="@string/app_name"
        android:theme="@style/AppTheme" >
        <activity
            android:name=".activity.MainActivity"
            android:label="@string/app_name" >
            <intent-filter>
                <action android:name="android.intent.action.MAIN" />
                <category
    android:name="android.intent.category.LAUNCHER" />
            </intent-filter>
        </activity>
        <activity
            android:name=".activity.AddElectricCarActivity"
            android:parentActivityName=".activity.MainActivity"
            android:label="@string/app_name" >
        </activity>
        <activity
            android:name=".activity.ShowElectricCarActivity"
            android:parentActivityName=".activity.MainActivity"
            android:label="@string/app_name" >
        </activity>

        <provider
            android:name=".provider.ElectricCarContentProvider"
            android:authorities="com.example.contentproviderdemo1"
```

```
                    android:enabled="true"
                    android:exported="true"
                    android:readPermission="com.example.permission.
➡READ_ELECTRIC_CARS"
                    android:writePermission="com.example.permission.
➡WRITE_ELECTRIC_CARS">
        </provider>
    </application>
</manifest>
```

Pay special attention to the lines in bold. Under **<application>** there are declarations of three activities and a provider. There are also two **permission** elements that define the permissions that external applications need to request to access the content provider.

The content provider, represented by the **ElectricCarContentProvider** class, is shown in Listing 27.2. Note the static final **CONTENT_URI** that defines the URI for the provider. Note also that **ElectricCarContentProvider** uses a database manager that takes care of data access and manipulation.

Listing 27.2: The content provider

```java
package com.example.contentproviderdemo1.provider;
import android.content.ContentProvider;
import android.content.ContentUris;
import android.content.ContentValues;
import android.database.Cursor;
import android.net.Uri;
import android.util.Log;

public class ElectricCarContentProvider extends ContentProvider {

    public static final Uri CONTENT_URI =
            Uri.parse("content://com.example.contentproviderdemo1"
                    + "/electric_cars");

    public ElectricCarContentProvider() {
    }

    @Override
    public int delete(Uri uri, String selection,
                        String[] selectionArgs) {
        String id = uri.getPathSegments().get(1);
        return dbMgr.deleteElectricCar(id);
    }

    @Override
    public String getType(Uri uri) {
        throw new UnsupportedOperationException("Not implemented");
    }

    @Override
    public Uri insert(Uri uri, ContentValues values) {
        long id = getDatabaseManager().addElectricCar(values);
        return ContentUris.withAppendedId(CONTENT_URI, id);
```

```
    }

    @Override
    public boolean onCreate() {
        // initialize content provider on startup
        // for this example, nothing to do
        return true;
    }

    @Override
    public Cursor query(Uri uri, String[] projection,
                        String selection,
                        String[] selectionArgs,
                        String sortOrder) {
        if (uri.equals(CONTENT URI)) {
            return getDatabaseManager()
                    .getElectricCarsCursor(projection, selection,
                            selectionArgs, sortOrder);
        } else {
            return null;
        }
    }

    @Override
    public int update(Uri uri, ContentValues values,
                        String selection,
                        String[] selectionArgs) {
        String id = uri.getPathSegments().get(1);
        Log.d("provider", "update in CP. uri:" + uri);
        DatabaseManager databaseManager = getDatabaseManager();
        String make = values.getAsString("make");
        String model = values.getAsString("model");
        return databaseManager.updateElectricCar(id, make, model);
    }

    private DatabaseManager dbMgr;
    private DatabaseManager getDatabaseManager() {
        if (dbMgr == null) {
            dbMgr = new DatabaseManager(getContext());
        }
        return dbMgr;
    }
}
```

ElectricCarContentProvider extends **ContentProvider** and overrides all its abstract
methods for CRUD operations. At the end of the class there is a definition of
DatabaseManager and a method named **getDatabaseManager** that returns a
DatabaseManager. The **DatabaseManager** is presented in Listing 27.3. It is similar to
the **DatabaseManager** class discussed in Chapter 18, "Working with the Database"
which explains how it works in detail. Please refer to this chapter if you have forgotten
how to work with relational databases.

Listing 27.3: The database manager

```java
package com.example.contentproviderdemo1.provider;
import android.content.ContentValues;
import android.content.Context;
import android.database.Cursor;
import android.database.sqlite.SQLiteDatabase;
import android.database.sqlite.SQLiteOpenHelper;
import android.util.Log;

public class DatabaseManager extends SQLiteOpenHelper {
    public static final String TABLE_NAME = "electric_cars";
    public static final String ID_FIELD = "_id";
    public static final String MAKE_FIELD = "make";
    public static final String MODEL_FIELD = "model";
    public DatabaseManager(Context context) {
        super(context,
                /*db name=*/ "vehicles_db",
                /*cursorFactory=*/ null,
                /*db version=*/1);
    }
    @Override
    public void onCreate(SQLiteDatabase db) {
        String sql = "CREATE TABLE " + TABLE_NAME
                + " (" + ID_FIELD + " INTEGER, "
                + MAKE_FIELD + " TEXT,"
                + MODEL_FIELD + " TEXT,"
                + " PRIMARY KEY (" + ID_FIELD + "));";
        db.execSQL(sql);

    }

    @Override
    public void onUpgrade(SQLiteDatabase db, int arg1,
                int arg2) {
        db.execSQL("DROP TABLE IF EXISTS " + TABLE_NAME);
        // re-create the table
        onCreate(db);
    }

    public long addElectricCar(ContentValues values) {
        Log.d("db", "addElectricCar");
        SQLiteDatabase db = this.getWritableDatabase();
        return db.insert(TABLE_NAME, null, values);
    }

    // Obtains single ElectricCar
    ContentValues getElectricCar(long id) {
        SQLiteDatabase db = this.getReadableDatabase();
        Cursor cursor = db.query(TABLE_NAME, new String[] {
                        ID_FIELD, MAKE_FIELD, MODEL_FIELD},
                        ID_FIELD + "=?",
                new String[] { String.valueOf(id) }, null,
```

```
                    null, null, null);
        if (cursor != null) {
            cursor.moveToFirst();
            ContentValues values = new ContentValues();
            values.put("id", cursor.getLong(0));
            values.put("make", cursor.getString(1));
            values.put("model", cursor.getString(2));
            return values;
        }
        return null;
    }

    public Cursor getElectricCarsCursor(String[] projection,
            String selection,
            String[] selectionArgs, String sortOrder) {
        SQLiteDatabase db = this.getReadableDatabase();
        Log.d("provider:" , "projection:" + projection);
        Log.d("provider:" , "selection:" + selection);
        Log.d("provider:" , "selArgs:" + selectionArgs);
        return db.query(TABLE_NAME, projection,
                selection,
                selectionArgs,
                sortOrder,
                null, null, null);
    }

    public int updateElectricCar(String id, String make,
            String model) {
        SQLiteDatabase db = this.getWritableDatabase();
        ContentValues values = new ContentValues();
        values.put(MAKE_FIELD, make);
        values.put(MODEL_FIELD, model);
        return db.update(TABLE_NAME, values, ID_FIELD + " = ?",
                new String[] { id });
    }

    public int deleteElectricCar(String id) {
        SQLiteDatabase db = this.getWritableDatabase();
        return db.delete(TABLE_NAME, ID_FIELD + " = ?",
                new String[] { id });
    }
}
```

The **CONTENT_URI** in **ElectricCarContentProvider** specifies the URI used for accessing the content provider. However, client applications should only know the content of this URI and do not need to depend on this class. The **Util** class in Listing 27.4 contains a copy of the URI for the clients of the content provider.

Listing 27.4: The Util class

```
package com.example.contentproviderdemo1;
import android.net.Uri;
public class Util {
    public static final Uri CONTENT_URI =
```

```
            Uri.parse("content://com.example.contentproviderdemo1" +
                    "/electric_cars");
    public static final String ID_FIELD = "_id";
    public static final String MAKE_FIELD = "make";
    public static final String MODEL_FIELD = "model";
```

}

Listings 27.5, 27.6 and 27.7 are activity classes that access the content provider. They all access the content provider by using the **ContentResolver** object created for the application. You retrieve it by calling getContentResolver from the activity classes.

Listing 27.5: The MainActivity class

```
package com.example.contentproviderdemo1.activity;
import android.app.Activity;
import android.content.Intent;
import android.database.Cursor;
import android.os.Bundle;
import android.view.Menu;
import android.view.MenuItem;
import android.view.View;
import android.widget.AdapterView;
import android.widget.AdapterView.OnItemClickListener;
import android.widget.CursorAdapter;
import android.widget.ListAdapter;
import android.widget.ListView;
import android.widget.SimpleCursorAdapter;
import com.example.contentproviderdemo1.R;
import com.example.contentproviderdemo1.Util;

public class MainActivity extends Activity {

    @Override
    protected void onCreate(Bundle savedInstanceState) {
        super.onCreate(savedInstanceState);
        setContentView(R.layout.activity_main);
        ListView listView = (ListView) findViewById(
                R.id.listView);
        Cursor cursor = getContentResolver().query(
                Util.CONTENT_URI,
                /*projection=*/ new String[] {
                        Util.ID_FIELD, Util.MAKE_FIELD,
                        Util.MODEL_FIELD},
                /*selection=*/ null,
                /*selectionArgs=*/ null,
                /*sortOrder=*/ "make");
        startManagingCursor(cursor);
        ListAdapter adapter = new SimpleCursorAdapter(
                this,
                android.R.layout.two_line_list_item,
                cursor,
                new String[] {Util.MAKE_FIELD,
                        Util.MODEL_FIELD},
```

```
                    new int[] {android.R.id.text1, android.R.id.text2},
                    CursorAdapter.FLAG_REGISTER_CONTENT_OBSERVER);

        listView.setAdapter(adapter);
        listView.setChoiceMode(ListView.CHOICE_MODE_SINGLE);
        listView.setOnItemClickListener(
                new OnItemClickListener() {
                    @Override
                    public void onItemClick(
                            AdapterView<?> adapterView,
                            View view, int position, long id) {
                        Intent intent = new Intent(
                                getApplicationContext(),
                                ShowElectricCarActivity.class);
                        intent.putExtra("id", id);
                        startActivity(intent);
                    }
                });
    }

    @Override
    public boolean onCreateOptionsMenu(Menu menu) {
        getMenuInflater().inflate(R.menu.menu_main, menu);
        return true;
    }

    @Override
    public boolean onOptionsItemSelected(MenuItem item) {
        switch (item.getItemId()) {
            case R.id.action_add:
                startActivity(new Intent(this,
                        AddElectricCarActivity.class));
                return true;
            default:
                return super.onOptionsItemSelected(item);
        }
    }
}
```

Listing 27.6: The AddElectricCarActivity class

```
package com.example.contentproviderdemo1.activity;
import android.app.Activity;
import android.content.ContentValues;
import android.os.Bundle;
import android.view.Menu;
import android.view.View;
import android.widget.EditText;
import com.example.contentproviderdemo1.provider.
➥ElectricCarContentProvider;
import com.example.contentproviderdemo1.R;

public class AddElectricCarActivity extends Activity {
```

```
    @Override
    protected void onCreate(Bundle savedInstanceState) {
        super.onCreate(savedInstanceState);
        setContentView(R.layout.activity_add_electric_car);
    }

    @Override
    public boolean onCreateOptionsMenu(Menu menu) {
        getMenuInflater().inflate(R.menu.add_electric_car, menu);
        return true;
    }

    public void cancel(View view) {
        finish();
    }

    public void addElectricCar(View view) {
        String make = ((EditText) findViewById(
                R.id.make)).getText().toString();
        String model = ((EditText) findViewById(
                R.id.model)).getText().toString();
        ContentValues values = new ContentValues();
        values.put("make", make);
        values.put("model", model);
        getContentResolver().insert(
                ElectricCarContentProvider.CONTENT_URI, values);
        finish();
    }
}
```

Listing 27.7: The ShowElectricCarActivity class

```
package com.example.contentproviderdemo1.activity;
import android.app.Activity;
import android.app.AlertDialog;
import android.content.ContentUris;
import android.content.ContentValues;
import android.content.DialogInterface;
import android.database.Cursor;
import android.net.Uri;
import android.os.Bundle;
import android.util.Log;
import android.view.Menu;
import android.view.MenuItem;
import android.view.View;
import android.widget.EditText;
import android.widget.TextView;
import com.example.contentproviderdemo1.R;
import com.example.contentproviderdemo1.Util;

public class ShowElectricCarActivity extends Activity {
    long electricCarId;
```

```java
@Override
protected void onCreate(Bundle savedInstanceState) {
    super.onCreate(savedInstanceState);
    setContentView(R.layout.activity_show_electric_car);
    getActionBar().setDisplayHomeAsUpEnabled(true);
    Bundle extras = getIntent().getExtras();
    if (extras != null) {
        electricCarId = extras.getLong("id");
        Cursor cursor = getContentResolver().query(
                Util.CONTENT_URI,
                /*projection=*/ new String[] {
                        Util.ID_FIELD, Util.MAKE_FIELD,
                        Util.MODEL_FIELD},
                /*selection=*/ " id=?",
                /*selectionArgs*/ new String[] {
                        Long.toString(electricCarId)},
                /*sortOrder*/ null);
        if (cursor != null && cursor.moveToFirst()) {
            String make = cursor.getString(1);
            String model = cursor.getString(2);
            ((TextView) findViewById(R.id.make))
                    .setText(make);
            ((TextView) findViewById(R.id.model))
                    .setText(model);
        }
    }
}

@Override
public boolean onCreateOptionsMenu(Menu menu) {
    getMenuInflater().inflate(R.menu.show_electric_car, menu);
    return true;
}

@Override
public boolean onOptionsItemSelected(MenuItem item) {
    switch (item.getItemId()) {
        case R.id.action_delete:
            deleteElectricCar();
            return true;
        default:
            return super.onOptionsItemSelected(item);
    }
}

private void deleteElectricCar() {
    new AlertDialog.Builder(this)
        .setTitle("Please confirm")
        .setMessage(
                "Are you sure you want to delete " +
                        "this electric car?")
        .setPositiveButton("Yes",
            new DialogInterface.OnClickListener() {
```

```
            public void onClick(
                    DialogInterface dialog,
                    int whichButton) {
                Uri uri = ContentUris.withAppendedId(
                    Util.CONTENT_URI, electricCarId);
                getContentResolver().delete(
                    uri, null, null);
                dialog.dismiss();
                finish();
            }
        })
    .setNegativeButton("No",
            new DialogInterface.OnClickListener() {
                public void onClick(
                        DialogInterface dialog,
                        int which) {
                    dialog.dismiss();
                }
            })
    .create()
    .show();
}

public void updateElectricCar(View view) {
    Uri uri = ContentUris.withAppendedId(Util.CONTENT_URI,
            electricCarId);
    ContentValues values = new ContentValues();
    values.put(Util.MAKE_FIELD,
            ((EditText)findViewById(R.id.make)).getText()
                    .toString());
    values.put(Util.MODEL_FIELD,
            ((EditText)findViewById(R.id.model)).getText()
                    .toString());
    getContentResolver().update(uri, values, null, null);
    finish();
}
}
```

As the content provider is accessed from components in the same application, you should not expect to encounter any problems. Figure 27.1 shows a **ListView** in the main activity. Of course, when you first run the application, the list will be empty.

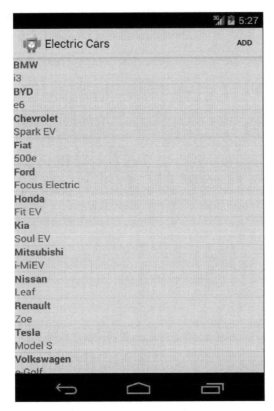

Figure 27.1: The main activity

Touch the Add button on the action bar to add an electric car. Figure 27.2 shows how the Add activity looks like.

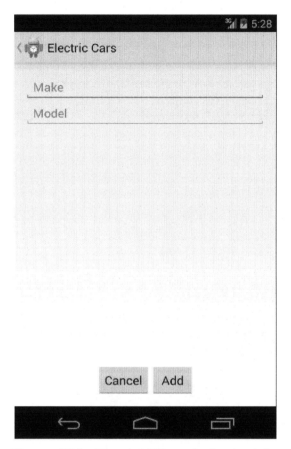

Figure 27.2: The AddElectricCarActivity

Type in a make and a model and touch the Add button to add a vehicle. Alternatively, touch the Cancel button to cancel. You will be redirected to the main activity.

From the main activity, you can select a car to view and edit the details. Figure 27.3 shows the **ShowElectricCarActivity** activity.

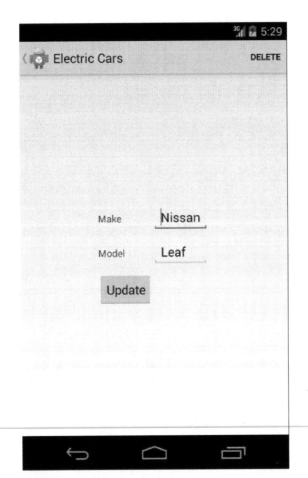

Figure 27.3: The ShowElectricCarActivity

You can update a car or delete it from this activity.

Consuming A Content Provider

The second application in this chapter, the ContentResolverDemo1 project, shows how you can access a content provider from a different application. The only difference between accessing a content provider from the same application and an external application is that you have to request a permission to access the provider in the manifest of the external application.

Listing 27.8 shows the manifest for the ContentResolverDemo1 project.

Listing 27.8: The manifest for ContentResolverDemo1

```xml
<?xml version="1.0" encoding="utf-8"?>
<manifest xmlns:android="http://schemas.android.com/apk/res/android"
    package="com.example.contentresolverdemo1" >

    <uses-permission
       android:name="com.example.permission.READ_ELECTRIC_CARS"/>
    <uses-permission
       android:name="com.example.permission.WRITE_ELECTRIC_CARS"/>

    <application
        android:allowBackup="true"
        android:icon="@drawable/ic launcher"
        android:label="@string/app name"
        android:theme="@style/AppTheme" >
        <activity
android:name="com.example.contentresolverdemo1.MainActivity"
            android:label="@string/app_name" >
            <intent-filter>
                <action android:name="android.intent.action.MAIN" />

                <category android:name="android.intent.category.LAUNCHER"
    />
            </intent-filter>
        </activity>
    </application>

</manifest>
```

The application contains one activity that shows data from the content provider. The activity class is a copy of the MainActivity class in the ContentProviderDemo1 project. The activity is shown in Figure 27.4.

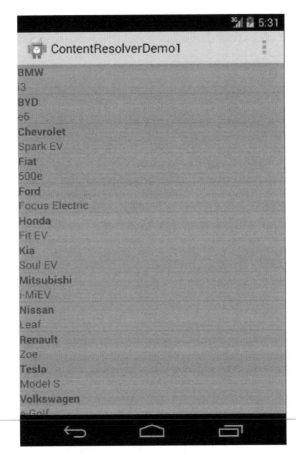

Figure 27.4: Showing data from a content provider

Summary

A content provider is an Android component used for encapsulating data that is to be shared with other applications. This chapter shows how you can create a content provider and consume its data from an external application using a **ContentResolver**.

Appendix A
Installing the JDK

You need the Java SE Development Kit (JDK) to create Android applications. This appendix shows you how to download and install it.

Downloading and Installing the JDK

Before you can start compiling and running your programs, you need to download and install the JDK as well as configure some system environment variables. You can download the latest version of the JDK for Windows, Linux, and Mac OS X from this Oracle website:

```
http://www.oracle.com/technetwork/java/javase/downloads/index.html
```

If you click the Download link on the page, you'll be redirected to a page that lets you select an installation for your platform: Windows, Linux, Solaris or Mac OS X. The same link also provides the JRE. However, for development you need the JDK not only the JRE, which is only good for running compiled Java classes. The JDK includes the JRE.

After downloading the JDK, you need to install it. Installation varies from one operating system to another. These subsections detail the installation process.

Installing on Windows

Installing on Windows is easy. Simply double-click the executable file in you downloaded in Windows Explorer and follow the instructions. Figure A.1 shows the first dialog of the installation wizard.

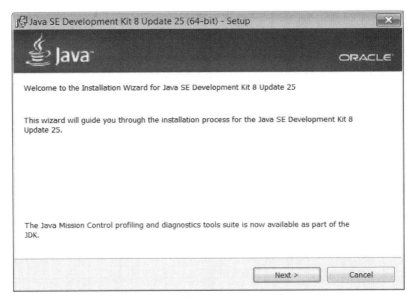

Figure A.1: Installing the JDK 8 on Windows

Installing on Linux

On Linux platforms, the JDK is available in two installation formats.

- RPM, for Linux platforms that supports the RPM package management system, such as Red Hat and SuSE.
- Self-extracting package. A compressed file containing packages to be installed.

If you are using the RPM, follow these steps:

1. Become root by using the **su** command
2. Extract the downloaded file.
3. Change directory to where the downloaded file is located and type:

```
chmod a+x rpmFile
```

where *rpmFile* is the RPM file.

4. Run the RPM file:

```
./rpmFile
```

If you are using the self-extracting binary installation, follow these steps.

1. Extract the downloaded file.
2. Use **chmod** to give the file the execute permissions:

```
chmod a+x binFile
```

Here, *binFile* is the downloaded bin file for your platform.
3. Change directory to the location where you want the files to be installed.
4. Run the self-extracting binary. Execute the downloaded file with the path prepended to it. For example, if the file is in the current directory, prepend it with

" ./":

./binFile

Installing on Mac OS X

To install the JDK 8 on Mac OS X, you need an Intel-based computer running OS X 10.8 (Mountain Lion) or later. You also need administrator privileges. Installation is straight-forward.

1. Double-click on the .dmg file you downloaded.
2. In the Finder window that appears, double-click the package icon.
3. On the first window that appears, click **Continue**.
4. The Installation Type window appears. Click **Install**.
5. A window appears that says "Installer is trying to install new software. Type your password to allow this." Enter your Admin password.
6. Click **Install Software** to start the installation.

Setting System Environment Variables

After you install the JDK, you can start compiling and running Java programs. However, you can only invoke the compiler and the JRE from the location of the **javac** and **java** programs or by including the installation path in your command. To make compiling and running programs easier, it is important that you set the **PATH** environment variable on your computer so that you can invoke **javac** and **java** from any directory.

Setting the Path Environment Variable on Windows

To set the **PATH** environment variable on Windows, do these steps:

1. Click **Start**, **Settings**, **Control Panel**.
2. Double-click **System**.
3. Select the **Advanced** tab and then click on **Environment Variables**.
4. Locate the **Path** environment variable in the **User Variables** or **System Variables** panes. The value of **Path** is a series of directories separated by semicolons. Now, add the full path to the **bin** directory of your Java installation directory to the end of the existing value of **Path**. The directory looks something like:

   ```
   C:\Program Files\Java\jdk1.8.0_<version>\bin
   ```

5. Click **Set**, **OK**, or **Apply**.

Setting the Path Environment Variable on UNIX and Linux

Setting the path environment variable on these operating systems depends on the shell you use. For the C shell, add the following to the end of your ~/**.cshrc** file:

```
set path=(path/to/jdk/bin $path)
```

where *path/to/jdk/bin* is the bin directory under your JDK installation directory.

For the Bourne Again shell, add this line to the end of your ~/**.bashrc** or ~/**.bash_profile** file:

```
export PATH=/path/to/jdk/bin:$PATH
```

Here, *path/to/jdk/bin* is the **bin** directory under your JDK installation directory.

Testing the Installation

To confirm that you have installed the JDK correctly, type **javac** on the command line from any directory on your machine. If you see instructions on how to correctly run **javac**, then you have successfully installed it. On the other hand, if you can only run **javac** from the **bin** directory of the JDK installation directory, your **PATH** environment variable was not configured properly.

Downloading Java API Documentation

When programming Java, you will invariably use classes from the core libraries. Even seasoned programmers look up the documentation for those libraries when they are coding. Therefore, you should download the documentation from here.

```
http://www.oracle.com/technetwork/java/javase/downloads/index.html
```

(You need to scroll down until you see "Java SE 8 Documentation.")

The API is also available online here:

```
http://download.oracle.com/javase/8/docs/api
```

Appendix B
Using the ADT Bundle

This appendix shows how you can create an Android application using the ADT Bundle. It also explains how to setup an emulator so you can develop, test, debug, and run Android applications even if you do not have a real Android device.

Installing the ADT

If you already have Eclipse on your local machine, you can install the ADT plug-in only and work with your existing Eclipse. However, note that it is easier to install the ADT bundle. If you choose to install the ADT plug-in, information on how to proceed with it can be found here.

```
http://developer.android.com/sdk/installing/installing-adt.html
```

To install the ADT Bundle, first download the ADT bundle from this site.

```
http://developer.android.com/sdk/index.html
```

Unpack the downloaded package to your workspace. The main directory will contain two folders, **eclipse** and **sdk**. Navigate to the **eclipse** folder and double-click the Eclipse program to start it. You will be asked to select a workspace. After that, the Eclipse IDE will open. The main window is shown in Figure B.1. Note that the application icon of ADT Eclipse is different from that of "regular" Eclipse.

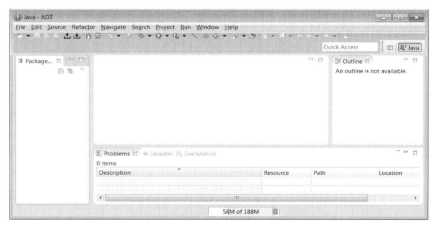

Figure B.1: The ADT window

Now you are ready to write your first Android application.

Creating An Application

Creating an Android application with the ADT Bundle is as easy as a few mouse clicks. This section shows how to create a Hello World application, package it, and run it on an emulator. Make sure you have installed the ADT Bundle by following the instructions in Introduction.

Next, follow these steps.

1. Click the New menu in Eclipse and select **Android Application Project**. Note that in this book Eclipse refers to the version of Eclipse included in the ADT Bundle or Eclipse with the ADT plug-in installed. The **New Android Application** window will open as shown in Figure B.2.

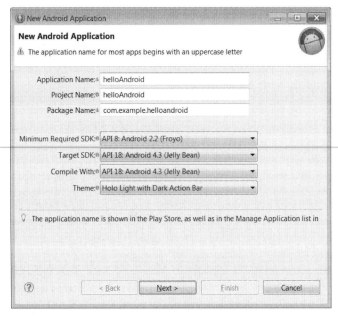

Figure B.2: The New Android Application window

2. Type in the details of the new application. In the **Application Name** field, enter the name you want your application to appear on the Android device. In the **Project Name** field, type a name for your project. This can be the same as the application name or a different name. Then, enter a Java package name in the **Package Name** field. The package name will uniquely identify your application. Even though you can use any string that qualifies as a Java package, the package name should be your domain name in reverse order. For example, if your domain name is example.com, your package name should be **com.example**, followed by the project name.

Now, right under the text boxes are four dropdown boxes. The **Minimum Required SDK** dropdown contains a list of Android SDK levels. The lower the level, the more devices your application can run on, but the fewer APIs and features you can use. The **Target SDK** box should be given the highest API level your application will be developed and tested against. The **Compile With** dropdown should contain the target API to compile your code against. Finally, the **Theme** dropdown should contain a theme for your application.

For your first application, use the same values as those shown in Figure B.2.

3. Click **Next**. You will see a window similar to the one in Figure B.3. Accept the default settings.

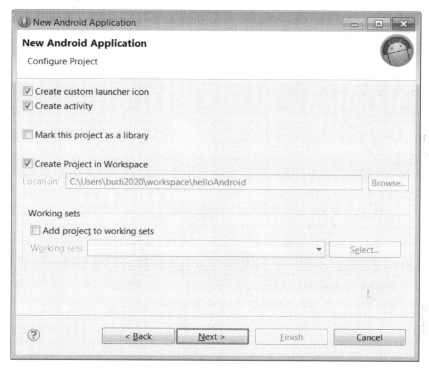

Figure B.3: Configuring your application

4. Click **Next** again. The next window that appears will look like the window in Figure B.4. Here you can choose an icon for your application. If you don't like the default image icon, click **Clipart** and select one from the list. In addition, you can use text as your icon if you so wish.

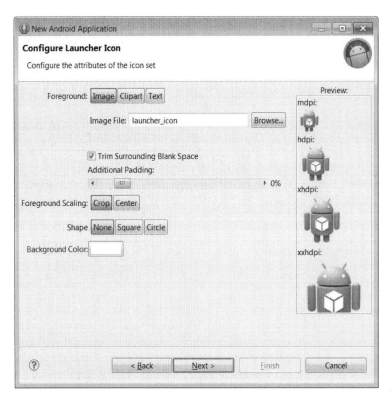

Figure B.4: Selecting a launcher icon

5. Click **Next** again and you will be prompted to select an activity (See Figure B.5). Leave **Blank Activity** selected.

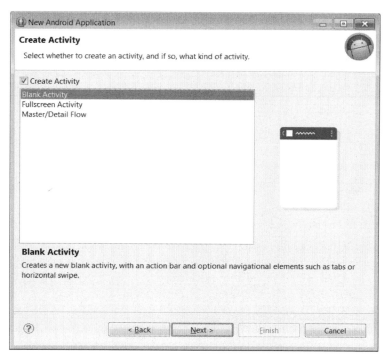

Figure B.5: Selecting an activity type

6. Click **Next** one more time. The next window will appear as shown in Figure B.6.

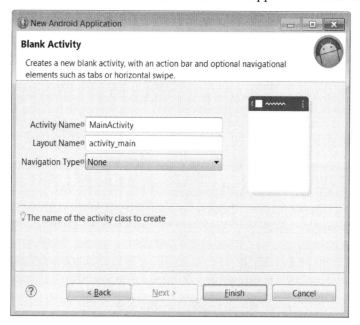

Figure B.6: Entering the activity and layout names

7. Accept the suggested activity and layout names and click **Finish**. The ADT Bundle

will create your application and you'll see your project like the screenshot in Figure B.7.

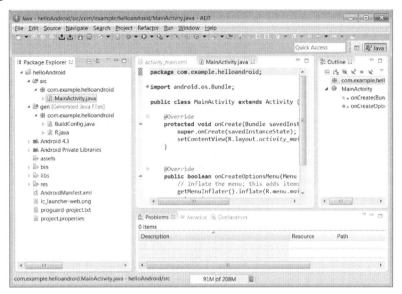

Figure B.7: The new Android project

In the root directory of Eclipse's Package Explorer (on the left), you'll find the following files:

- **AndroidManifest.xml** file. This is an XML document that describes your application.
- An icon file in PNG format.
- A **project.properties** file that specifies the Android target API level.

On top of that, there are the following folders.

- **src**. This is your source code folder.
- **gen**. This is where generated Java classes are kept. The generated Java classes allow your Java source to use values defined in the layout file and other resource files. You should not edit generated files yourself.
- **bin**. This is where the project build will be saved in. The application APK will also be found here after you have run your application successfully.
- **libs**. Contains Android library files.
- **res**. Contains resource files. Underneath this directory are these directories: **drawable-xxx** (containing images for various screen resolutions), **layout** (containing layout files), **menu** (containing menu files), and **values** (containing string and other values).

One of the advantages of developing Android applications with an IDE, such as ADT Eclipse, it knows when you add a resource under the res directory and responds by updating the **R** generated class so that you can easily load the resource from your program. You will learn this powerful feature in the chapters to come.

Running An Application on An Emulator

The ADT Bundle comes with an emulator to run your applications on if you don't have a real device. The following are the steps for running your application on an emulator.

1. Click the Android project on the Eclipse Project Explorer, then click **Run > Run As > Android Application**.
2. The **Android Device Chooser** window will pop up (see Figure B.8). (Once you configure it, it will not appear the next time you try to run your application).

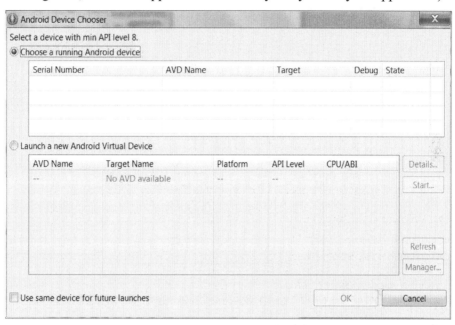

Figure B.8: The Android Device Chooser window

3. Here you can choose to run your application on a real Android device (an Android phone or tablet) or an Android Virtual Device (emulator). In Figure B.8 you do not see a running Android device because no real device is connected, so click the **Launch a new Android Virtual Device** radio button, and click the **Manager** button on the right. The **Android Virtual Device Manager** window will appear (See Figure B.9).

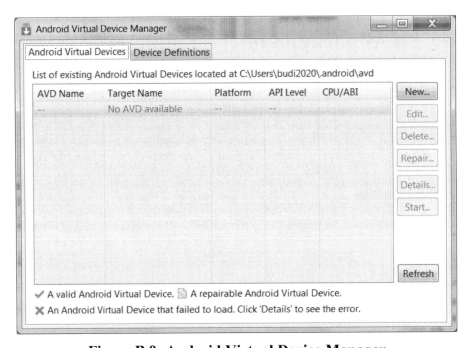

Figure B.9: Android Virtual Device Manager

4. Click **New** on the **Android Virtual Devices** pane to display the **Create new AVD** windows (See Figure B.10)

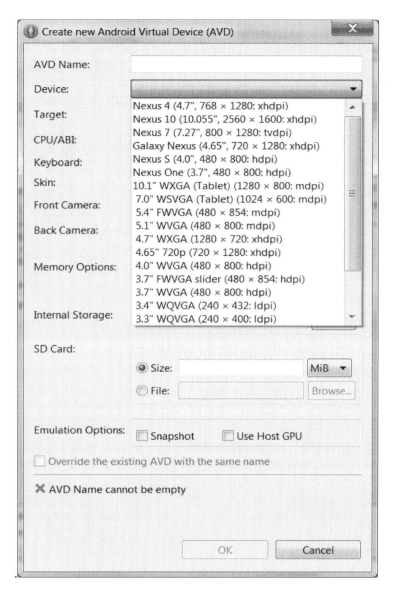

Figure B.10: Creating a new virtual device

5. Click the **Device** drop-down to view the list of virtual devices available. Here I choose Nexus 7. Then, give your device a name. The name must not contain spaces or any special characters.
6. Choose a target and if you're using Windows, reduce the RAM to 768. For some reason, it may crash if you're using more than 768MB RAM on Windows.
7. My options are shown in the screenshot in Figure B.11.

Figure B.11: Entering values for a new virtual device

8. Click **OK**. The **Create new Android Virtual Device (AVD)** window will close and you'll be back at the **Android Virtual Device Manager** window. Your AVD will be listed there, as shown in Figure B.12.

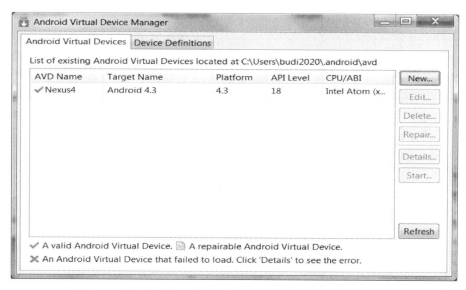

Figure B.12: The list of virtual devices available

9. Now, click the AVD name (Nexus7) to select it and the **Start** and other buttons will be enabled. Click the **Start** button to start the AVD. You will see the Launch Options popup like that in Figure B.13.

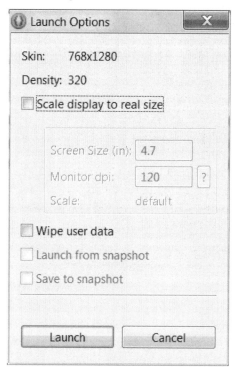

Figure B.13: The Launch Options popup

10. Click **Launch** to launch your AVD. You'll see a window like that in Figure B.14

when it's launching.

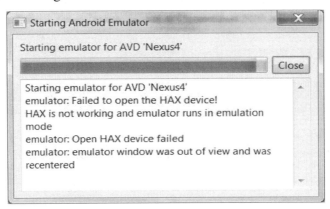

Figure B.14: Starting the emulator

It will take a few minutes or more depending on your computer speed. Figure B.15 shows the emulator when it is ready.

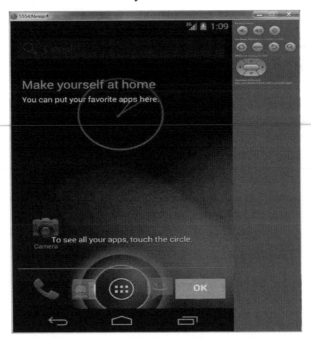

Figure B.15: The Android emulator

As you know, the emulator emulates an Android device. You need to unlock the screen by touching (or clicking) the blue circle at the bottom.
If your application does not open automatically, locate the application icon and double-click on it. Figure B.16 shows the HelloWorld application.

Figure B.16: Your first application on the emulator

During development, leave the emulator running while you edit your code. This way, the emulator does not need to be loaded again every time you run your application.

Logging

Java programmers like to use logging utilities, such as Commons Logging and Log4J, to log messages. The Android framework provides the **android.util.Log** class for the same purpose. The **Log** class comes with methods to log messages at different log levels. The method names are short: **d** (debug), **i** (info), **v** (verbose), **w** (warning), **e** (error), and **wtf** (what a terrible failure).

This methods allow you to write a tag and the text. For example,

```
Log.e("activity", "Something went wrong");
```

During development, messages logged using the **Log** class will appear in the LogCat view in Eclipse. If you don't see it, click **Window → Show View → LogCat** or **Window → Show View → Other → LogCat**.

The good thing about LogCat is that messages at different log levels are displayed in different colors. In addition, each message has a tag and this makes it easy to find a message. In addition, LogCat allows you to save messages to a file and filter the messages so only messages of interest to you are visible.

The LogCat view is shown in Figure B.17.

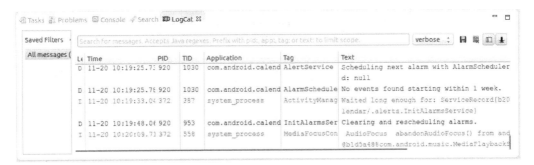

Figure B.17: The LogCat view

Any runtime exception thrown, including the stack trace, will also be shown in LogCat, so you can easily identify which line of code is causing the problem.

Debugging An Application

Even though Android applications do not run on the JVM, debugging an Android application in Eclipse does not feel that different from debugging Java applications.

The easiest way to debug an application is by printing messages using the **Log** class. However, if this does not help and you need to trace your application, you can use the debugging tools in Android.

Try adding a line break point in your code by double-clicking the bar to the left of the code editor. Figure B.18 shows a line breakpoint in the code editor.

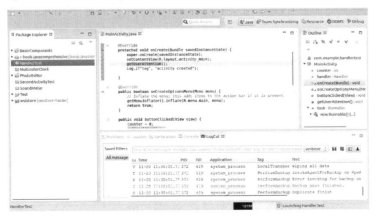

Figure B.18: A line breakpoint

Now, debug your application by clicking the project icon in the Project Explorer and selecting **Run → Debug As → Android Application**.

Eclipse will display a dialog asking you whether you want to open the Debug perspective. Click **Yes**, and you will see the Debug perspective like the one in Figure B.19.

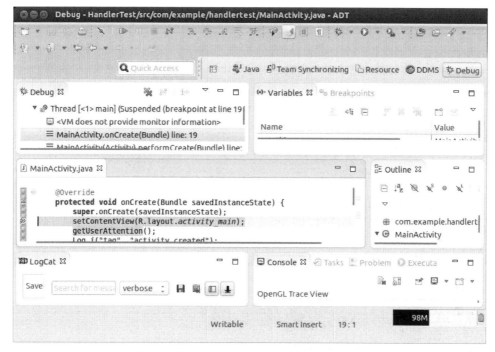

Figure B.19: The Debug perspective

Here, you can step into your code, view your variables, and so on.

In addition to a debugger, Android also ships with Dalvik Debug Monitor Server (DDMS), which consists of a set of debugging tools. You can display the DDMS in Eclipse by showing the DDMS perspective. (See Figure B.20).

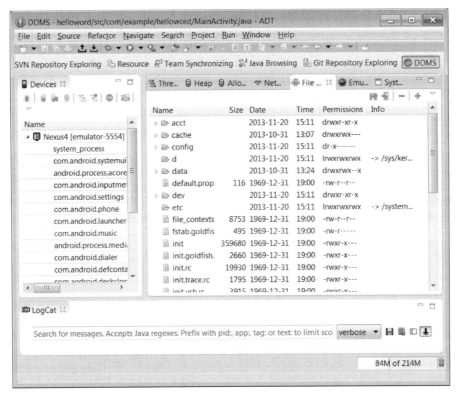

Figure B.20: The DDMS perspective in Eclipse

You will see LogCat as one of the views in the DDMS perspective. However, you can also use DDMS to do any of these:

- Verify that a device is connected.
- View heap usage for a process
- Check object memory allocation
- Browse the file system on a device
- Examine thread information
- Monitor network traffic

Index